CRASH!

Military Aircraft Disasters, Accidents and Incidents

CRASH!

Military Aircraft Disasters, Accidents and Incidents

ANDREW BROOKES

LONDON

IAN ALLAN LTD

First published 1991

ISBN 0 7110 1965 7

Published by Ian Allan Ltd, Shepperton, Surrey;
and printed by Ian Allan Printing Ltd at their works
at Coombelands in Runnymede, England

To my wife Teresa

Front cover:
**The trickiest runways to operate into and
out of are those, like Gibraltar and Kai Tak,
Hong Kong, which stretch into the sea. It is
not just that the ever-approaching watery
vista can concentrate the mind wonderfully,
but also there is no margin for error on
either side if things go wrong. A Canberra
PR7, on detachment from Singapore to Kai
Tak, found this out when it broke its back on
veering off the runway.**

Contents

Introduction

Flying is not inherently dangerous; statistically, it is nearly 20 times safer to get airborne in an airliner than to travel by car. It is just that the sky can be a particularly unforgiving place to be. Whether you sit in an air defence fighter or a basic trainer, whether you exceed the speed of sound or transit leisurely at 50mph, when it all goes quiet there is no pulling into the kerb and calling a garage. If you hit a motorway barrier, you may be deflected off; there is no bouncing a Harrier off a hillside. The sky is an alien environment where lift is a gift but thrust is a must — as Michael Flanders used to say, 'if God had intended us to fly, He would never have given us the railways'.

Yet man has always courted danger and for modern military man, aerial combat offers a chance to show individual daring and initiative far removed from any perceived contribution he can make to the massed armies or fleets down below. The dilemma comes in balancing all that elan with prudence such that a modern two-seat jet fighter — which may cost well over £10 million, not counting £3 million for training the crew — is not written off while practising to fight effectively on the morrow.

All military flying necessarily includes an element of risk, and a balance has always to be struck between flight safety considerations and acceptance of those risks which are essential to complete the task. This book therefore sets out to describe some military flying accidents that have happened over the years, not only to be informative but also to help strike the right balance between operating and surviving. However, it is worth making the point at the outset that the military is unlikely to eradicate flying accidents completely. The only way to guarantee that is never to fly, which negates the whole purpose of acquiring an expensive air force in the first place.

In dealing with past accidents, I have tried to pass on food for thought rather than pious platitudes. In the process, I hope that I do not give any offence to friends or relatives of the deceased aviators involved. I regard all those who had the misfortune to lose their lives in tragic circumstances as colleagues who would wish successive generations to learn from their experiences. This book is certainly not written to point the finger at anyone — I have enough RAF flying hours under my belt to admit that sometimes, there but for the Grace of God went I.

It is interesting to note that the terminology and songs of military aircrew for nearly 80 years have dealt overwhelmingly with death and dismemberment of men and machines. Yet 'crash' is a word I have rarely heard used to describe what happens when several tons of metal ploughs itself and its crew into the ground. A World War 2 American flier would say that someone 'bought the farm', and his RAF equivalent would talk of 'prangs'. Today an RAF man might mention in passing that so-and-so has 'gone in', or an American refer to a 'one-way trip to Harp City', but whatever the synonym it is usually delivered with a measured flippancy that acts as a defence mechanism. Back in 1912, the oldest squadron song of all began a trend that essentially expressed a need to get fear out of the system:

The Bold Aviator   or   The Dying Airman

Oh, the bold aviator was dying,
And as 'neath the wreckage he lay, he lay,
To the sobbing mechanics about him
These last parting words he did say:

Chorus:
Two valve springs you'll find in my stomach,
Three spark plugs are safe in my lung, my lung,
The prop is in splinters inside me,
To my fingers the joy-stick has clung.

Oh, get you two little white tombstones,
Put them one at my head and my toe, my toe,
And get you a pen-knife and scratch there,
'Here lies a poor pilot below'.

Take the propeller boss out of my liver,
Take the aileron out of my thigh, my thigh,
From the seat of my pants take the piston,
Then see if the old crate will fly.

And when at the Court of Inquiry
They ask for the reason I died, I died,
Please say I forgot twice iota
Was the minimum angle of glide. Oh —

Take the cylinders out of my kidneys
Take connecting rod out of my brain, my brain
From the small of my back take the crankshaft,
And assemble the en-gyne again.

And when I join the Air Force
Way, way up in the sky, the sky,
Let's hope that they know twice iota
Is the minimum angle to fly. Oh —

Take the cylinders out of my kidneys
Take connecting rod out of my brain, my brain,
From the small of my back take the crankshaft,
And assemble the en-gyne again.

7

If fatalism was understandable in the pioneering age of flimsy flying machines held together by little more than bracing wire and hope, modern aircraft designed by computers, filled with fail-safe features and having maintainability and reliability built-in from birth, are much less likely to stop working completely or come apart in the air. Nevertheless the human dimension stays constant and I occasionally have the feeling that there is an emerging generation of aircrew, brought up in the video age, who are very much at home with modern technology but who lack that basic respect for the hazards of flying that their predecessors once expressed in verse and banter. At the end of the day, situational awareness in the air is no different in a Tornado from that in a Lancaster or Sopwith Camel — only the time frame for sorting things out may vary. If this book stops anyone from thinking that aircraft cannot bite, and stimulates thought as to what to do should that awful moment occur, it will have done its job.

Below:
A typical through-canopy zero speed, ground level test being conducted with a Martin-Baker Mk 10 lightweight ejection seat. *Martin-Baker*

The First Accident

If the Wright brothers had been a bit more superstitious, they might have had reservations about choosing the Kill Devil Hills, sandwiched in the Outer Banks of North Carolina near the village of Kitty Hawk, as an ideal site for experiments in soaring flight. But choose it they did, and on the morning of 17 December 1903 Wilbur Wright achieved the first powered, controlled and sustained aerodynamic flight by man. It was a far cry from their original bicycle business, but even at this early stage Orville and Wilbur prophetically concluded on the train back home to Dayton for Christmas that their infant creation held great potential as a weapon of war.

The year 1904 was one of consolidation as the brothers concentrated on improving their 'Flyer' and its controllability: in fact their total flying time for the year amounted to only 45min. However, in 1904 and 1905 they were actively involved in promoting their creation as well as developing it. By 1905 the British War Office was indicating a definite interest, with the French not far behind, but at home officialdom remained distinctly underwhelmed such that by the end of the year the Wrights decided to end their flying altogether. Stung by press scepticism and scorn, angered by their own government's rejection, and convinced that almost everyone with whom they came into contact was out to exploit or cheat them, they withdrew into their shells. The years 1906 and 1907 were spent in further research, and the brothers' luck only turned the following year when both the US Army Signal Corps and the French offered contracts.

On 8 February 1908 the Signal Corps officially sought a machine 'capable of carrying two men and sufficient fuel supplies for a flight of 125 miles, with a top speed of at least 40 miles an hour'. More power was given to the 1905 'Flyer' together with upright seats, and Wilbur flew the modified two-seater for the first time on 6 May 1908.

Having now to meet firm customer deadlines for the first time, the brothers went hard at their flying. During one period of nine days, they flew every day for up to 15min at a time, which may not seem all that wonderful today but it must be remembered that they never flew at all between November 1905 and May 1908. On one of these flights, Wilbur let the controls get away from him and over-controlled with the front elevator. The aeroplane was flying downwind at about 50mph when, without warning, it departed from controlled flight and crashed into the sand. Wilbur emerged from the grinding crash with a bruised cheek, some hurt ribs, and a solid blow across his nose.

In subsequent years, it would become axiomatic that any pilot who crashed was sent back up into the air straight away to recover his nerve. The Wrights could afford no such luxury; they only had one flying machine. Thus, on a personal level, they had to wait days or even weeks, mulling over what went wrong and perhaps worrying about what might go wrong the next time. But it was of just as much concern that Wilbur's crash forced the brothers to make major decisions. Time is precious in aviation, and time was running swiftly away from the Wrights. Because of their contract, they had planned to build a new 1908 'Flyer' for demonstration to the Signal Corps at Fort Myer, Virginia, in September. If it was not ready, the entire contract would be in jeopardy. As pressure rarely comes in dribs and drabs, the French simultaneously announced that they too would not countenance any delays in meeting their needs, and that the Wrights must prove their worth almost immediately in France. While Orville returned to Ohio to begin work on the new 'Flyer', Wilbur set sail across the Atlantic: the brothers who had worked as one for so long were now to be separated in their flying endeavours for the first time in their lives.

Wilbur took France by storm, and his name was soon to sweep the entire continent. In the meantime, his younger brother worked away on the new machine such that by 3 September 1908 he was ready for his first test. Yet when he arrived at Fort Myer, he saw that a group of Army officers and a small throng of spectators had collected. It was obvious that most were convinced that Orville was intending to start and complete his entire test at one go. Despite not having flown for a long time, and being faced with a first flight in an untried machine from a postage stamp of a field, Orville found his first experimental hop becoming a circuit immediately after take-off. The age of public pressure on pilots had arrived.

In France, 18 September dawned bright and beautiful. At 08.00hrs Wilbur prepared his machine at the Le Mans racetrack for what was intended to be a record-breaking flight, only to receive a telegram that Orville had crashed. The accident had happened about 17.00hrs on the day before, Thursday 17 September. Orville had been circling the field at Fort Myer with everything going well when suddenly the aircraft pitched over and dove straight at the ground. Orville survived the crash with a fractured left leg, four broken ribs, a fractured and dislocated hip, and back injury, but his official army passenger, Lt Thomas E. Selfridge, suffered a fractured skull and died one hour after reaching hospital. He was the first man to die in the crash of a powered aircraft.

It took two days before a shocked Wilbur felt able to write to his sister:

'I received the news of the awful accident at Washington only on Friday morning. The death of poor Selfridge was a greater shock to me than Orville's injuries, severe as the latter were. I felt sure that Orville would pull through all right, but the other was irremediable.

'I cannot help thinking over and over again, "If only I had been there, it would not have happened".

'It was not right to leave Orville to undertake such a task alone. I do not mean that Orville was incompetent to do the work itself, but I realised that he would be surrounded by thousands of people who with the most friendly intentions in the world would consume his time, exhaust his strength, and keep him from having proper rest. When a man is in this condition he tends to trust more to the carefulness of others instead of doing everything and examining everything himself. A man cannot take sufficient care when he is subject to constant interruptions and his time is consumed in talking to visitors. I cannot help suspecting that Orville told others to put on the big screws instead of doing it himself, and that if he had done it himself he would have noticed that thing that made the trouble, whatever it may have been. If I had been there I could have held off the visitors while he worked or let him hold them off while I worked. But he had no one to perform this service. People think I am foolish because I do not like the men to do the least important work on the machine. They say I crawl under the machine and over the machine when the men could do the thing well enough. I do it partly because it gives me the opportunity to glance around to see if anything in the neighbourhood is out of order. Hired men pay no attention to anything but the particular thing they are told to do, and are blind to everything else. It is much easier to do things when you have someone at hand in whom you have absolute confidence.'

From Dayton a convalescing Orville wrote this accident report on 14 November to Wilbur in France:

'It is two weeks today since I left the hospital at Fort Myer, yet I am just beginning to get about the house on crutches. This is the first I have written since the accident.

'We had made three rounds of the ground, keeping well inside of buildings, trees, etc, so that the turns were of necessity pretty short. On the fourth round, everything seemingly working much better and smoother than in any former flight, I started on a larger circuit with less abrupt turns. It was on the very first slow turn that the trouble began. Just after passing over the top of our buildings at a height which I estimate at 100 or 110ft, and while travelling directly towards Arlington Cemetery, I heard (or felt) a light tapping in the rear of the machine. A hurried glance behind revealed nothing wrong, but I decided to shut off the power and descend as soon as the machine could be faced in a direction where a landing could be made. This decision was hardly reached when two big thumps, which gave the machine a terrible shaking, showed that something had broken. At the time I only thought of the transmission. The machine suddenly turned to the right and I immediately shut off the power. I then discovered that the machine would not respond to the steering and lateral balancing levers, which produced the most peculiar feeling of helplessness. Yet I continued to push the levers, when the machine suddenly turned (the right wing rising high in the air) till it faced directly up the field. I reversed the levers to stop the turning and to bring the wings on a level. Quick as a flash, the machine turned down in front and started straight for the ground. Our course for 50ft was within a few degrees of the perpendicular. Lt Selfridge up to this time had not uttered a word, though he took a hasty glance behind when the propeller broke, and turned once or twice to look into my face, evidently to see what I thought of the situation. But when the machine turned headfirst for the ground, he exclaimed, "Oh! Oh!" in an almost inaudible voice.

'I pulled the front rudder lever to its limit, but there was no response in the course of the machine. Thinking that, maybe, something was caught and that the rudder was not completely turned, I released the lever a little and gave another pull, but there was no change. I then looked at the rudder and saw that it was bent to its limit downward, and that the pressure of the air on the under side was bulging the cloth up between the ribs. The first 50ft of that plunge seemed like a half-minute, though I can hardly believe that it was over one second at most. The front rudder in that distance had not changed the course

11

more than five or 10 degrees. Suddenly just before reaching the ground, probably 25ft, something changed — the machine began to right itself rapidly. A few feet more, and we would have landed safely. As it was, the skids hit out at the front end. All the front framing was broken and the machine turned up on edge.

'The only explanation I have been able to work out of the cause of the plunge for the ground is that the rear rudder, after the stay wire was torn loose by the propeller, fell over on its side and in some mysterious manner was caught and held in this position, with a pressure on its under side . . .'

After every military accident, especially one involving fatalities, the appropriate military commander tasks an independent team with carrying out an in-depth inquiry into the causes. Back in 1908, the same rules applied and the Chief Signal Officer, US Army, tasked an Aeronautical Board composed of Maj Saltzmann, Capt Wallace and Lt Lahm to provide a detailed report on the accident. They took testimony from soldiers who witnessed the event, interviewed Orville, and sought expert advice from distinguished engineers such as Octave Chanute.

Their final report, dated 19 February 1909, was a model of brevity and clarity. The Board found that Mr Wright and Lt Selfridge took off at 17.14hrs to circle the field to the left as usual. They completed 4½ turns and had been airborne for 4min 18sec when witnesses heard a report and then saw a section of propeller blade flutter to the ground. The machine was judged to be at 150ft, it then glided down for perhaps 75ft. 'At this point it appeared to stop, turn so as to head up the field toward the hospital, rock like a ship in rough water, then drop straight to the ground in the remaining 75ft.' The Board concluded that the accident 'was due to the accidental breaking of a propeller blade and a consequent loss of control which resulted in the machine falling to the ground'. The fact that Tom Selfridge was the heaviest passenger Orville had taken up was irrelevant, and the unfortunate lieutenant, who had been very anxious to make the flight before leaving for St Joseph, Missouri, to fly Dirigible No 1, was buried with full military honours at Arlington Cemetery.

Nevertheless it has to be said that back in 1908, the only informed authority capable of investigating meaningfully into the Wright brothers' activities were the Wright brothers themselves. In the middle of 1909, and after thinking long and hard about the accident, Wilbur outlined his conclusions on the cause in a letter to Octave Chanute:

'After looking over the Fort Myer machine we have decided that the trouble came in the following manner. One blade of the right propeller developed a longitudinal crack which permitted the blade to flatten out and lose its pushing power. The opposite blade, not being balanced by an equal pressure on the injured blade, put strains on its axle and its supports which permitted it to swing forward and sideways a little farther than the normal position and at the

Right:
Orville Wright (in flat hat) and Lt Selfridge prepare to get airborne on their ill-fated flight from Fort Myer, Va, on 17 September 1908. *Library of Congress*

same time set up a strong vibration. This brought the uninjured blade in contact with the upper stay wire to the tail and tore it loose, the end of the wire wrapping round the end of the blade and breaking it off. The blade which broke was not the one which originated the trouble. The machine was pointing almost towards Arlington Cemetery at this moment, but swerved to the right where small trees made a bad landing place. Orville in the meantime had stopped the engine and attempted to turn to the left so as to land in the regular grounds, but found the tail inoperative. He therefore twisted the wings so as to present the left wing at a greater angle in order that the increased resistance might hold that end of the machine back and face the machine up the field. The manoeuvre succeeded very well and the machine had faced back and descended a third or more of the distance to the ground without any indication of serious trouble. He next moved the lever to straighten the wing tips so as to go straight ahead, but the machine instantly turned down in front and made almost straight for the ground. He thinks the tail had fallen over on its side with possibly a slight negative angle and that, when he moved the handle which operates the wings and tail, the latter was twisted to a positive angle and received a pressure on the under side, which caused the plunge. He pulled the front rudder to the limit, but for a time the response was very slow. Toward the end something seemed to change and the machine began to right, but it was too late. The ground was

struck at a very steep angle and with terrific speed. The splitting of the propeller was the occasion of the accident, the uncontrollability of the tail was the cause . . .'

Some press headlines — 'Wright Plane Falls, Army Lieutenant, a Passenger, Killed' — reflected popular realisation of the fact that flying could seriously damage your health. What troubled Wilbur, apart from bitter self-criticism that he had left undone something that could have prevented the crash, was that the public might turn against everything he and his brother were striving for because of what had happened at Fort Myer. But the reaction never came and if anything, public enthusiasm and support became even greater than before. In London, *The Times* spoke for all thinking men when its editorial accepted human flight as something to be regarded as a normal feature of the world's future life.

So on Monday 21 September 1908, with the Sabbath behind him, Wilbur flew over France for 1hr 31min and 25sec and set a new world record distance of 41 miles. Thereafter, he took to the air time and again like a man possessed to carry the banner for aircraft in general and Orville in particular. On 3 October he and a passenger flew for nearly 35 miles, thereby gaining instant acceptance of his machine's ability to perform with more than a single occupant. Despite poor Tom Selfridge's accident, there was no turning back for military aviation, only a wholehearted commitment to getting it right the next time. Having proved that there was nothing inherently wrong with their creation, and that the best way of surviving in the air is to know your machine thoroughly, Wilbur Wright put the risks into perspective when he said in 1908: 'To be completely safe, you must sit on the fence and watch the birds'.

Lighter Than Air

There is a long tradition of portraying pilots as courageous, death-defying individuals whose heroic stature is measured by the degree of risk they are willing to take in the air. In the early days of aviation this emphasis on courage was perhaps inevitable, since so much of the theory and technique of flying was unknown or misunderstood and had to be learned by trial and error. Nevertheless, the wisest heads never played up the bravado of flying. Wilbur and Orville Wright, who concentrated on designing and building aeroplanes that were reasonably safe for the very good reason that the brothers wished to sell them at a profit, died in their beds from ailments in no way related to defying gravity.

The image of the pilot first as a solitary, courageous warrior, and then as leader of a crew bonded together in a spirit of derring-do to face all challenges, emerged from the role of aviation in World War 1. This tradition continued in one form or other through the record-setting and barn-storming 1920s, and it was given added impetus in military circles by the need to keep aviation in the public eye to fend off army and navy top brass bent on relegating aircraft to the sidelines or off the map altogether.

When the pilot as hero was not fighting the enemy, he was engaged in mortal combat with the winds, clouds, storms, oceans, ice, mountains and all the primordial forms of nature — not to mention the limits of human endurance. As these struggles were heavily romanticised by the press and cinema it was easy to lose sight of the fact that, even in military flying, the emphasis should always be on attention to detail and the avoidance of unnecessary risk.

Early development of the aeroplane was probably hindered because the aerodynamics involved were (and largely remain) a mystery to the man in the street. On the other hand, the mechanics of getting an airship to fly were obvious to all who had ever inflated a party balloon. Hydrogen is lighter than air; encase it in rubber and it will float upwards; hang on to it and like Winnie the Pooh you will rise, too.

From 1887 the electrolytic production of aluminium enabled a very lightweight rigid structure of girders to be built around hydrogen-filled gas bags. Add to this Herr Daimler's engines, a touch of Count von Zeppelin's genius, and the airship became a practical proposition. By 1914 Zeppelins had carried 37,250 passengers over 90,000 miles of scheduled air routes without death or injury to any of them. An airship passenger, relaxing in his wicker armchair sipping fine wine at a steady 45mph, could only look down in all respects on the relatively uncivilised and uncomfortable stutterings of heavier-than-air craft below.

Came World War 1 and out of the airship went all the agreeable cold lunches, armchairs, carpets and stewards to be replaced by bombs and guns. New skills were learned from 1915 onwards as German crews set course to bomb England at night. Nevertheless, on 7 June 1915 a parvenu heavier-than-air Morane Saulnier L at the hands of Sub-Lt R. A. J. Warneford bombed LZ 37 from *above*, bringing the airship down in flames across the convent of St Elizabeth near Ghent. Out of the Zeppelin's crew, there was only one survivor.

Thereafter German airship designers went for height to elude the defences. A new breed of 'height climbers', each 645ft long, entered service with the German Naval Airship Division in 1917, and their operating levels around 20,000ft placed them well above the range of Allied aircraft or guns. Yet very few advantages are conferred for nothing. German weather forecasters were quite unable to predict gales in the upper air, and accurate navigation was extremely difficult at previously unheard-of heights where landmarks were almost invisible if not obscured by clouds or darkness. Lower oxygen pressure above 10,000ft also led to altitude sickness, and hand-in-very-thick-glove was the extremely bitter cold which made stiff joints and frostbite the greatest enemy.

Their crews swathed in piles of extra clothing, 13 'height climbers' set out on 19 October 1917 with orders to 'attack Middle England industrial regions'. Yet despite their impressive number, five were destined never to return as unpredicted high level northerly gales blew all the ships astray. Faced with a rapid southerly drift, the force leader should have ordered his craft back before they reached England, but it was not to be. One of their number was L 49. Her English landfall was 100 miles in error and her bombs fell on open country. Three of her five engines failed, probably from the same oxygen starvation that affected her crew, so there was no means of sustaining the height that was the Zeppelin's prime means of salvation. Blown across France by the unexpected winds, the helpless L 49 was close behind L 44 which was shot down in flames by anti-aircraft fire over St Clement. Lost and demoralised, and with the radio that could have given bearings out of commission, Kapitänleutnant Gayer and his crew of 18 in L 49 turned back west in the belief that they were near Holland. At 6,500ft they were bounced by five Nieuport scouts. Frozen and exhausted, and completely lost, Gayer hoisted the white flag at 3,000ft before force-landing in the wooded slopes of the River Apance. Owing to a faulty Very pistol, Gayer was unable to set L 49 alight before he and his crew were disarmed by local farmers, hunters and four of the Nieuport pilots who had landed in adjacent fields.

Delighted at having captured a 'height climber' almost intact, Allied experts made detailed examinations of L 49. This was particularly good news for the British who were so far behind in airship technology that no rigid airship of their design had flown even as late as the middle of 1916. But once it was fired up, the British Admiralty made a conscious effort to surpass the opposition in size, payload, ceiling, range and speed. Constructor-Cdr Charles Campbell was confident that he could produce an all-British giant superior to any Zeppelin, and in June 1918 the Naval Staff formulated a requirement for a rigid airship able to patrol over the North Sea for six days up to 300 miles from base. The first of this class, R 38, was ordered from Short Brothers at Cardington in September 1918. The size of the ship had to be limited to fit inside Shorts' shed,

Right:
Zeppelin L 49, still intact but downed in woods on the slopes of the River Apance near Bourbonne-les-Baines, France, in October 1917.

but even after scaling down from the original concept R 38 was to be 700ft long, 180ft wide and 110ft high to carry 2,724,000cu ft of hydrogen in 14 gas cells. The new airship was heavily engined to reach a maximum speed of 62kt (70mph), it had to carry 30 tons of fuel for 65hr flight flat out, yet the hull was to be as light as possible to achieve a combat ceiling of 22,000ft. R 38 embodied many refinements from the downed L 49, such as lightened hull with lighter girders, but in his enthusiasm Campbell designed only half the strength and safety factors of previous ships into R 38. He was fatally ignorant of the fact that in turns, much higher bending loads were imposed on the hull than wind tunnel tests had so far indicated.

Work on the R 38 started around February 1919, but by this time the war had ended and financial pressures on the Treasury dictated that retrenchment and economy were to be the orders of the day. Across the Atlantic, a US Navy plan for four rigid airships was halved for similar reasons; one was now to be purchased abroad, the other to be built at home, and an airship base for both to be established at Lakehurst, New Jersey. Thus when the British cancelled many airships under construction, work continued on R 38 to enable it to be sold in October 1919 to the US Government for £400,000. R 38 was to be numbered

ZR-2 in America, pending design and construction in the US of the ZR-1. In the same month the Airship Dept was transferred from the Admiralty to the Air Ministry.

At this time the erection of the two R 38 main frames had just been completed, but in the spring of 1920 Campbell found himself appointed manager of the Royal Airship Works when Cardington was compulsorily taken over from Shorts. As early as May 1920, some Americans started to express doubts as to the R 38's structural integrity, but Campbell's administrative duties distracted him from his design efforts. His burden was not eased by having to make sea trips to the US to consult with the designers of the ZR-1. All of which may partially explain why, for instance, when the R 38's after engines were uprated with a subsequent weight increase of over 1,200lb, no effort was made to strengthen the ship's structure.

In March 1921 Flt Lt John Pritchard, the airship trials officer, proposed that R 38 should do 100hr of test flying in British hands, including rough weather flying, followed by 50hr with an American crew before the US Navy took R 38 across the Atlantic. Pritchard had qualms about the fragility of the lightly-built 'height climber', and he subsequently urged that R 38 be tested at a minimum altitude of 7,000ft where lower air density reduced aerodynamic loads. German 'height climbers' had been tested in this manner, but such considerations held little sway. The RAF's Director of Supply and Research, who was Campbell's superior in the Air Ministry, ignored the unprecedented dimensions and new features of R 38 and insisted on trials being completed in 50hr. He was supported by other non-technical officials who saw that much money could be saved if the airship was handed over at least six months earlier than expected. Cdr Lewis H. Maxfield, the designated US commander, also did not press for more than 50hrs of trials before he took over, and his crew who had been sent over to the UK for training were equally eager to return home as soon as possible in the largest and finest airship yet built.

Pritchard's misgivings were reinforced by R 38's first two trial flights in June when it was discovered that the control surfaces were overbalanced. The third flight in July transferred R 38 to the Howden airship base in East Yorkshire so that USN personnel based there could become familiar with it. While en route at 50kt, the airship began to hunt vertically over a range of 500ft. Pritchard took control of the wheel personally just as word came to him that transverse girders amidships had failed. This damage was caused by excessive high speed bending loads acting on the light structure, and it should have set warning bells ringing about R 38's airworthiness at her designated top speed. But no one in authority wanted to listen. Campbell immediately attributed the girder failures to slipstream from the port forward propeller beating on a large panel, rather than to compression caused by severe bending loads on the hull in the vertical plane. Perhaps Campbell fell into the very human trap of feeling that he had to justify his work, but his theory cut no ice with Pritchard who protested to Air Cdre Edward Maitland, AOC Howden, that it would be criminal to fly R 38 again without a thorough inspection. Maitland, an experienced aviator, tried to hold out for 150hr minimum trials, but even he was told to hand over the ship to the Americans as agreed by the Director of Supply and Research and not to give any more advice unless asked for it. Unfortunately the Director in question —

who had never flown more than twice in an airship — was away in Egypt, and the schedule had to go on, for to do otherwise would have cost money as well as loss of face.

If anyone could have called a halt it was the Americans, but Maxfield felt he was up against a British conspiracy to prevent him from commanding what he regarded as ZR-2 on its transatlantic flight. At a conference on 10 August, the Chief of Air Staff delicately expressed the Air Ministry's concern that Maxfield and his crew might be too inexperienced to undertake the crossing themselves. AM Trenchard's well-meaning offer to loan 'advisers' for the flight only engendered in Maxfield an unjustified mistrust of his British colleagues, and particularly of Pritchard, the only man who realised the limitations of the lightly built 'height-climber'.

Instead, after patching-up and some reinforcement, R 38 was sent on its final trials before making for the world's first mooring mast at Pulham, Norfolk, where the airship was to be loaded prior to its transatlantic flight to New Jersey. The tests were to include turns using 5° of rudder at 48kt, and on one occasion up to 10°.

On the morning of 23 August 1921, R 38 left Howden with a mixed crew of 28 Britons including Air Cdre Maitland, 17 Americans led by Maxfield, Cdr Campbell and four men from the National Physical Laboratory. Despite Maitland's presence, Flt Lt Archibald Wann was in command and at 07.10hrs R 38/ZR-2 left Howden. Half-an-hour later the sound of Sunbeam engines awakened the good people of Hull who flooded out into the streets to watch the great shape pass majestically through the early morning mist, circle, and head out to sea.

By 16.00hrs R 38 was 30 miles east of Lowestoft. It turned inland for a series of trials but as dusk fell, Wann found himself at 700ft in low cloud over Norfolk. With his officers unable to locate Pulham, he decided to stay airborne overnight and continue the trials the following day. Further engine and fuel consumption tests were run the next morning, but once it became clear that Pulham was going to stay fog-bound, it was decided to complete the trials and land back in Yorkshire. Navigating on radio bearings from Flamborough Head, R 38 descended from 3,000ft and flew up the Humber while signalling that she would carry out high speed tests and land around 18.30hrs.

With all six engines opened to 2,000rpm, a speed of 62kt was maintained for 10min before power was reduced. The ship handled well, and Campbell told Maitland that he was very pleased. Despite the power reduction, R 38 was still progressing at a very creditable 54½kt when Henry Bateman from the National Physical Laboratory finished photographing manometric equipment on the controls under the impression that the tests had ended. He was then told by Pritchard that there was to be a further test involving rapid movement of the rudders. What happened next will never be resolved. Wann, who was in the control car with Maitland, Pritchard and Maxfield, asserted later that he did not order the rudders put over by more than 15° on the helm indicator, and that with stretching of the control cables he did not believe that rudder travel was more than 10°. Furthermore, despite being moved from side to side, the rudders were briefly centred before being turned further. Bateman however, who was abaft of the rudders taking photographs, testified that the rudders at the end of the 10min turning test 'were being moved from hard over to hard over — a range of 50° altogether'. He also gained the impression that, 'an elevator test was being carried out at the same time'.

Down below in Hull, the shops and offices had just closed and for the second time in two days the populace gazed up in wonder at the beautiful sight of the huge silvery craft, her engines thundering, moving in and out of the clouds 2,000ft above. With no reduction in speed the rudders were put over to port, then to starboard, and back again, the angle increasing until they were being moved from hard left to hard right. As each turn became sharper, the aerodynamic loads on the rudders increased until the strain on the hull built up. Moreover the strains on the 700ft-long ship were much greater around 2,000ft than they would have been at the German testing level of 7,000ft, so it was almost inevitable that this reckless exercise would result in a major structural failure. At about 17.38hrs the airship, on a southeast heading, suddenly turned sharply west. Spectators were horrified to note a deepening diagonal wrinkle develop in the smooth outer cover on the starboard side 230ft from the stern. The nose and tail went down, the hull opened up at the top, and the ship broke its back between Frames 9 and 10. As the radio officer sent his final message, 'Ship broken — falling', the forward two-thirds of the rigid hull fell into the Humber river, rapidly engulfed in flames as a spark from the electrical leads met escaping fuel in the keel. A thunderous explosion rent the air as liberated hydrogen was ignited, followed by a second explosion as the forward portion hit the water. Windows were shattered over a two-mile radius of Hull, and although one woman died of shock, the casualty toll would have been much higher had the tons of wreckage not fallen into the river. As it was, only Flt Lt Wann, suffering shock and burns, survived from the front section.

The after third sunk more slowly, landing on a barely submerged sand bar directly off Corporation Pier. Four more survivors including Bateman were in this part which did not catch fire because all R 38's electrical sources were in the front portion. It was the fire that mainly accounted for 44 deaths out of the complement of 49, among whom were Maitland, Campbell, Pritchard, Maxfield and every American bar one.

By command of the Air Council, a court of inquiry led by AVM Sir John Salmond assembled at Howden at 10.00hrs on Saturday 27 August 'to inquire into the circumstances occasioning the loss of HM Airship R 38 and to express an opinion as to possible causes of the loss'. The court heard from many witnesses, including all the survivors apart from Wann, who was too ill to testify. The inquiry began by acknowledging that R 38 was designed 'to meet requirements which appear to the court to be greatly in advance of those of previous British airships. The requirements as to maximum height and speed, together with the limits in length imposed by the only available construction sheds, necessitated the utmost economy in hull weights and materials. Many new features were introduced in the design, and it appears evident that in some cases there was a lack of vital aerodynamic information as to the effect of these modifications on the strength of the structure.'

Given these factors, the archives were searched to determine if the Admiralty Board had ever approved the design of R 38, a standard requirement in warship production that went back more than a century. No trace of Board approval could be found, so it had to be concluded that Campbell's design had never been checked by any Crown authority. Nor was Campbell's work at Cardington ever subject to any independent check. The poor man was left to design and build the most advanced airship in the world, while running the Royal Airship Works that produced it, alone. The court concluded that such a state of affairs was 'unsound', which was a considerable understatement. Whatever Campbell's failings as a designer, he was a conscientious man who should never have been left to act as judge, jury and ultimate design authority for his own work. The court never strayed into muddier waters as to why Campbell was ever set such an impossible task in the first place. R 38 did not so much fail because of lack of theoretical information in Britain on aerodynamic stresses and loadings, but rather the airship could never have succeeded because it was designed around too many conflicting requirements.

After the Armistice there was no operational justification for maintaining a requirement for an airship to be built as light as possible so that it could attain 22,000ft with a warload. The Admiralty should have concentrated on what they really wanted, a low altitude North Sea long-range scout. In trying to combine two extremes — light enough to fly high plus strong enough to cope at low level with rough air and control surface loads while doing 60kt in co-operation with the Grand Fleet or sharply manoeuvring after a submarine — the Admiralty condemned Campbell to relying on a safety factor of four with respect to static loads and hoping that this would cover the aerodynamic ones as well. It was a piece of wishful thinking for which Campbell paid with his life.

Aircraft design always involves an element of compromise but, in their haste to explore the unknown, aircraft designers must never step so far beyond the limits of knowledge that they send aircrews off on little more than a wing and a prayer. The RAF court of inquiry obliquely sought to shuffle the blame for R 38's design failings on to the Admiralty, but the most junior service was not blameless. It was the Air Ministry which overstretched Campbell by adding administrative responsibility for Cardington to his not inconsiderable design burden. By the time of the accident they had also been responsible for R 38 for nearly two years, and with no grandiose airship dreams of their own to justify, the RAF should have recognised that extreme altitude was no longer essential and modified R 38 accordingly. The amount of material needed to double the strength of the longitudinal girders would have weighed four tons and would have lowered the R 38's ceiling by only 2,800ft.

All of which is obvious with hindsight, but the need for economy and pressure to meet the American contract all militated against changes and rushed completion of the ship. It is easy to point a finger at individuals; it is a lot more difficult for any one person to cry 'Halt!' when a great deal of national prestige is riding on a project.

Britain also fell down in clasping the wreck of the L 49 to its bosom without due consideration of how the Germans used her. Zeppelin 'height climbers' only operated at high speed in the thin upper air. They were handled very cautiously near sea level and their controls were specifically designed not to permit rapid movement of the rudders and elevators. If any German commander had handled a 'height climber' such as L 49 in the same way as R 38 was flown over the Humber, he would have been court martialled.

We will never know who made the fateful decision to carry out those extreme turning trials. The buckling on the third flight was the first significant indication of dangerous stresses, and it was only the prompt steadying of the ship by Flt Lt Pritchard that relieved them before the failure spread and broke up R 38 even earlier. Consequently, how could Pritchard, the airship's trials officer, have stood by and watched R 38's rudders being swung to their limits for 10min? Maitland had the authority to insist on such measures, but he had a reputation for never using his rank to intervene in the operation of any ship on which he flew. The important point though is not just that someone exceeded his pre-trial brief by a large margin, but also that a host of others in the control cabin stood by and let him do it. Apart from knowing the limits of his ship, a good captain always takes his crew along with his decisions. However, his authority must always be paramount. Flt Lt Wann must therefore take ultimate responsibility for the accident happening to R 38 when it did because he either exceeded his orders unilaterally or he let himself be overridden when the safety of his ship was at stake. By all means argue about it on the ground afterwards, but get safely on the ground first.

Having lost the bulk of their airship expertise in the R 38 accident, and in the process lost faith in British expertise and methods, the US Navy turned to German experience when it came to designing and building ZR-2's stablemate, the ZR-1. Starting out as a copy of the L 49 with American engines, ZR-1 was then lengthened, given more powerplants, and had her fins modified. She was also strengthened and, in the light of the R 38 accident, her design was checked by a competent and independent commission.

ZR-1 was to be the first rigid airship to fly with helium. About 90% of the world's supply of helium was concentrated in a small area around Amarillo, Texas, and the use of this non-inflammable gas was seen as vital given that explosions of hydrogen in the forward gas cells of R 38 had been responsible for much of the heavy death toll. But there were drawbacks, not least that an extra section had to be added to ZR-1 because helium gave less lift. Helium was also so rare and expensive that questionable precedures designed to conserve the precious gas came into use. The ship had to be flown excessively heavy to avoid discharging ballast, leading to severe bending loads on the hull. The number of automatic valves were also reduced and most dangerous of all, 'jampot' covers were placed on all automatic valves in flight.

The final ZR-1 design, approved on 31 October 1921, was for a rigid airship 680ft long, 78ft in diameter, and with a gas volume of 2,115,174cu ft in 20 gas cells. The airship first flew on 4 September 1923, and the following month she was christened *Shenandoah* — an Indian term meaning 'daughter of the stars' — by the wife of the Secretary of the Navy. Two years after the loss of ZR-2, it was a proud day for the US Navy. The *Shenandoah's* silvery beauty had been painstakingly researched such that the possibility of major structural failure in high speed, low level manoeuvres had been eliminated, she would be handled

Right:
USS *Shenandoah* (ZR-1) at her Lakehurst, NJ, mooring mast, circa 1923.
Admiral T. G. Kincaid, via US Naval Historical Center

with greater knowledge, and she would be lifted into the air by 'safe' helium. Surely nothing had been overlooked?

Shenandoah was designed to scout the great oceans around the USA but from the first moment she stopped traffic in the streets of New York and Washington, the admirals realised that they had a public relations winner on their hands. Even top brass who had no time for airships saw the value of one with 'US Navy' emblazoned on the side floating over the heads of voters who might live 1,000 miles from the coast.

So in between co-operation with the fleet, *Shenandoah* undertook a series of 'hand waving' publicity flights. All went well until the first week of September 1925 when the airship was scheduled for a Midwest flight to appear at five state fairs and 40 cities in six days. Her captain for the last 18 months had been Lt-Cdr Zachary Lansdowne and he was expected to adhere to a route and timings that had been publicised in advance. At 14.52hrs on 2 September, *Shenandoah* departed the mast at Lakehurst with 43 persons and 16,620lb of fuel on board for the first stage to Scott Field near St Louis.

Shenandoah crossed the Alleghenies by 01.45 the following day when, against a backdrop of lightning flashing ahead and to the north, Lansdowne set course for Zanesville, Ohio. He had known before he set off that there was a large low pressure area to the north of Minnesota, but this caused no undue concern to the crew of a ship that was exactly two years old and which had already flown

Below:
Cdr Zachary Lansdowne laying out his course on the bridge of *Shenandoah*, circa 1924. *US Naval Historical Center*

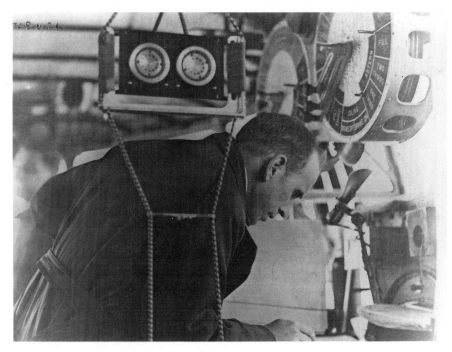

25,000 miles in all kinds of weather including storms. But although the air remained smooth, it soon became clear that a stiffening headwind from the southwest was markedly retarding *Shenandoah's* progress: changes in heading, level or engine speed made no difference.

At about 04.20hrs, being then at a height of 1,600ft, the elevator man suddenly cried a warning that the ship was starting to rise at 200ft/min and that he could not hold her. Lansdowne ordered the elevators down and the engines to full power, but even when nose-down by 18° the ship continued to rise at an average rate of 225ft/min. At 3,100ft the ascent was checked, but the *Shenandoah* was now close to her 'pressure height' of 3,600ft above which the thinner air would allow the helium to distend and totally fill the sealed bags, causing the airship to expand and tear herself to pieces. The order therefore went out to rip the jampot covers off the automatic valves to release the precious helium.

The nose-down angle had been too much for the lubrication system of two of the Packard engines and they failed. As the air became turbulent, the ship started pitching and rolling. After about 6min, *Shenandoah* began rising more rapidly than before despite being held down at an angle approaching 25°. A massive flow of cold air from the northwest, overrunning warm air flowing from the southwest, abruptly created severe instability over a wide area of Ohio and the *Shenandoah* was in a vertical current of air that was rising at 2,100ft/min. Helpless in the grip of such a force, she rapidly passed through her pressure height and Lansdowne, recognising that the eight automatic valves could not vent the expanding helium fast enough, ordered the manoeuvring valves opened as well. It was to little avail because even after 5mins' venting, the ship rose from 4,100ft to 5,360ft. When the valves were closed, the airship continued to rise even more rapidly to a peak of 6,060ft at a rate of nearly 1,000ft/min.

Sounds were now heard from the upper part of the hull of wires breaking and even girders collapsing. Suddenly a blast of cold air through the keel hatches showed that the ship had reached the overrunning cold air mass. Having lost 8,600lb of lift through venting helium, and despite having released 2,500lb of water ballast, the ship was 7,300lb heavy and she fell back towards the farmlands of Ohio as rapidly as she went up. Only the discharge of another 4,300lb of water ballast enabled her descent to be arrested at just under 3,000ft by another ascending air current.

All the way down the gas cells, their helium contracting, were flapping, rapidly deflating and probably becoming damaged. Realising that his ship might again be carried upward, and that only 2,500lb of water ballast remained, Lansdowne sent word to stand by to drop fuel tanks. As Lt-Cdr Charles Rosendahl went up to see that the keel watch was ready to drop, 'the ship took a very sudden upward inclination which seemed to me very much the same condition that exists in a plane upon the beginning of a loop'. A violent gust, striking the underside of the bow from starboard, forced the whole ship upward and rolled it over to port. Combined with a terrific vibration, the force was more than the structure could tolerate. At 04.52hrs, and after being carried up to 3,600ft, there was a loud crashing high in the ship on the port side as the giant hull broke in two over Ava, Ohio.

Shenandoah broke at Frame 125, about a third of the way from the front. For a minute-and-a-half, drifting with a 20mph northeasterly wind, the two halves of

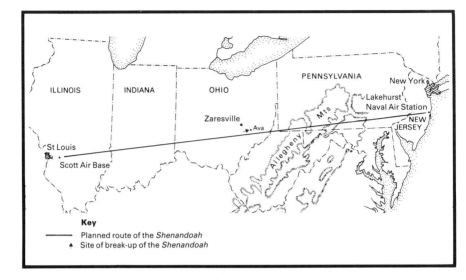

Key
——— Planned route of the *Shenandoah*
⬥ Site of break-up of the *Shenandoah*

the broken hull were held together by the massive four rudder and four elevator steel cables leading from the control car to the tail. But the tremendous pull exerted by the after portion of the broken hull soon had its effect. The control car was pulled aft by the cables with such force that its four occupants — including Lansdowne — were flung out.

With the final separation of the hull, a further break occurred at Frame 110 hurling two engine cars and their mechanics to the ground. The after portion of the hull fell rapidly but after the loss of the two engine cars, it levelled off and descended more slowly such that 18 men survived uninjured in the stern section.

As dawn broke, the 210ft bow section of the *Shenandoah*, now lightened of the weight of the control car, ascended possibly as high as 10,000ft. Briefly there were eight people on board but the ship's engineering officer, Lt Sheppard, had been standing at the break of Frame 125. He managed to grasp some wreckage as he fell, and when an Aviation Machinist's Mate tried to reach him he was heard to say, 'never mind me — look for yourself!' just before the structure to which he was clinging came free under his weight. Lt Sheppard's body, still clutching a piece of girder in one hand and a wire in the other, was found later in woods close to the wrecked tail section.

Finding that there was 1,600lb of water ballast still available, and that a manoeuvring valve could still be operated, Lt-Cdr Rosendahl and the other six survivors in the bow section felt confident that they could free-balloon to a landing. It must have been an amazing sight for any Ohio farmer to see 210ft of truncated dirigible dancing around the grey sky. In the best traditions of the service, Rosendahl shouted orders above the wind to his crew who, shaken and dizzy from their whirling motion, clung tightly to broken girders while ropes were dropped in the hope that they would catch on the ground. The first landing had to be aborted because of the high 25mph groundspeed, but not before the bow hit a walnut tree whose branches pulled one man out. Half-a-mile further on, the helium was valved off, gas cells were slashed open as high as the men could climb, and the trail ropes caught and held the bow on a hill near Sharon, Ohio at 05.45hrs.

On 14 September a court of inquiry headed by Rear-Adm Hilary P. Jones, a former CinC US Fleet, was appointed to investigate the loss of the *Shenandoah*. Lt-Cdr Rosendahl, the senior survivor, was first witness followed by all the others apart from the seriously injured man who had been plucked out by the walnut tree. One point was quickly resolved: it had been suggested that the loss of the control car led to the breakup of the ship, but it became clear from the witnesses' testimony that the car only tore away after the hull broke in two. Accusations were thrown around that the lack of sufficient automatic valves contributed to the loss of the ship, and that the *Shenandoah* was too underpowered to outrun the storm. It was even suggested that the girders of the hull were weakened in advance by corrosion. The fact that the flamboyant Brig-Gen Billy Mitchell was to be court martialled for his outspoken criticisms of some of the politics behind USN airship procurement was just one manifestation of the axe-grinding undertaken in and around the court of inquiry.

However, as far as the technical advisers to the court were concerned, 'it appears that the principal cause of the breaking was a great sagging moment at frame 120-130, due to an upward gust, directed obliquely from the starboard towards the port, and due to the speed of the ship through the air, at a considerable angle to the transverse plane'. The *Shenandoah* got into this trouble because she fell foul of a high velocity vertical current resulting from a steep temperature gradient between warm air at low altitude and cold air above 6,000ft. Such a climatic feature was characteristic of continental land masses crossed by mountain ranges, and it has to be said that this phenomena was little understood in 1925. In contrast, European airships had only previously operated in a maritime environment where cool winds off the sea and relatively little heating of the air over adjacent coastal areas were the norm. Some would lay the blame on the Americans for building the *Shenandoah* along basically European lines with insufficient thought to the weather of the continental US, but that misses the point. *Shenandoah*, unlike R 38, was well designed to cope with the rigours of low altitude scouting over the sea and she should have stuck to that. It was her employment on 'hand waving' missions hundreds of miles from the coast that brought about her demise, and the court did not hesitate to criticise the airship's employment on publicity flights:

'While recognising the propriety and necessity of the legally constituted authorities in the naval service being the sole judges of the sufficiency of the reasons actuating all orders to naval craft, and further recognising that the practice of ordering movements of naval vessels for the purpose of complying with public requests is in accord with long-established customs, it is considered that such movements should be limited to essentially naval and military operations in so far as possible, especially in the case of new and experimental types.'

In the end, the court exonerated Lansdowne by finding that, 'the final destruction of the ship was due primarily to large, unbalanced, external, aerodynamic forces arising from high velocity currents'. Many good recommendations were made for improving future airship design, such as making hulls fatter and thereby stronger in resisting bending forces, and integrating control cars into the keel structure. But the loss of *Shenandoah* following after R 38 started a public perception that the airship was an expensive deathtrap. The loss of the British R 101 in 1930, and the flaming *Hindenburg* disaster in 1937, only added weight to the arguments of heavier-than-air men who believed that the millions spent on airships could be better employed on more flexibly employable aircraft carriers.

Reading across from these airship accidents, aircraft should never be used for a purpose or in a fashion totally opposed to those for which they were designed,

28

without very detailed assessment of the consequences first. Increasingly sophisticated modern aircraft should also never blind anyone to the awesome power of nature. Good pre-flight and en route Met briefs are essential, and weather should always be treated with respect because it can still kill just as easily today as it did in 1925.

Finally, one of the survivors from the *Shenandoah* was Col Hall, an Army observer who had managed to scramble out of the control car just before it fell away. He closed his testimony to the court with the following tribute to Lansdowne and his men:

'While I was in the control car there was perfect order and discipline and the severe duties of the officers and enlisted men were performed as calmly and efficiently as in normal flight . . . I am sure that Captain Lansdowne and his officers and men in the control car went to their death attempting to keep control of the ship, which they had not lost until the actual breaking occurred.'

Lansdowne had flown 605.5hr on command of the *Shenandoah* and his inspired leadership, superb ship handling, and sixth sense about all the arcane variables associated with operating his airship, had stood him in good stead until a disaster arose with which no man could cope. Add the efforts of Lt-Cdr Rosendahl and his team to Lansdowne's record, and it becomes clear why men should never be replaced on aircraft flightdecks. When you are up to your armpits in alligators, the warning captions are flashing, the weather is foul and the reassuring radio is on the blink, the well-disciplined and knowledgeable brain is all that stands between salvation and disaster. As Winston Churchill put it in typically apposite fashion: 'to every man, there comes in his lifetime that special moment when he is tapped on the shoulder and offered the chance to do a very special thing, unique and fitted to his talents. What a tragedy if that moment finds him unprepared and unqualified for the work that would have been his finest hour.' No man is lost until he gives himself up as lost, but the man who takes to the air unsure of his craft and unprepared for all eventualities risks losing before he starts.

Don't Push Your Luck

The core of any air force is its flying training machine. Point a man in the right direction and the safety battle is half won; set him the wrong example and he may never live to regret it.

In 1912 the British established the Royal Flying Corps (RFC) comprising a Military and a Naval Wing served by a Central Flying School (CFS) which opened at Upavon on 19 June. Given the rudimentary nature of flying training in those days, it is perhaps surprising that only two major accidents occurred during the first flying course, though neither pilot was hurt. The Avro 500 training aircraft entered service soon afterwards with an undercarriage sturdy enough to absorb the most ham-fisted landing, but at a time when engines did not even have throttles – and Avro 500's Gnome engine had to be 'blipped' on and off for control during landing – life cannot have been easy for either instructor or pupil. Capt J. D. B. Fulton, commander of CFS's Avro flight, used to throw his arms around a fellow instructor whenever one of his students went solo and say: 'Just let me know what he's doing, I can't bear to look'.

Yet the School maintained a fine safety record until three fatal accidents occurred during the fifth course. The first happened on 3 October 1913 when Maj G. C. Merrick was coming in to land in perfect weather. His Short biplane was seen to glide steeply from about 300ft when the aircraft suddenly plunged violently down, turned on its back, and flung out the pilot who was killed immediately. Almost certainly Merrick was not strapped in, probably because like many others he preferred to be free to make a hasty exit on the ground in case the aircraft caught fire. He must have slipped out of his seat during the steep glide, jamming the stick forward so that the biplane became inverted.

In March 1914, Lt H. F. Treeby was killed when he stalled during a badly-judged landing, and nine days later Capt C. P. Downer's BE2c broke up in the air. Examination of the wreckage showed that Downer probably lowered the nose at 2,000ft for a normal landing approach, the elevators jammed, and he pulled so desperately to get out of the ever-steepening dive that he bent the stick. His efforts succeeded but the violent reversal of the elevators imposed such a strain that the wings collapsed.

It took the demands of World War 1 to increase both the tempo of training and the number of flying schools, and from 1914 onwards the public began to see military pilots as dashing knights of the air, impervious to fear and flippant in the face of danger. A typical example was the famous prewar sporting pilot, Capt Gordon-Bell, whose stammer and permanent monocle only added to the image. Flying a Bristol Scout during the retreat from Mons, Gordon-Bell managed to land upside down in a tree. There he was discovered by a high-ranking Staff Officer who, after carefully examining the upside-down pilot, asked in cultured tones: 'Have you had an accident?' Came the infuriated reply, 'N-n-no, I always land like this, you d-d-damned fool.'

Maj Robert Smith-Barry was another prime aviation character who shortly after arriving in France with No 5 Squadron, crashed his BE8 when its engine failed. His observer was killed but Smith-Barry escaped with badly broken legs. He managed to flee from his sickbed in St Quentin, just before the Germans occupied the town, by summoning an old horse-drawn four-wheel cab in which he solemnly drove off in his pyjamas with his splinted legs on the front seat. Smith-Barry had a limp and used a stick for the rest of his life, but this neither hampered his flying nor his enthusiasm for making flying safe.

There was certainly much scope for improvement in RFC attitudes towards operating safely. For instance, in 1916 a group of barely trained No 29 Squadron pilots were sent to France in DH2s. These 'pusher' scouts were not easy to fly and a young lieutenant called Harold Balfour (later to become Lord Balfour and Under-Secretary of State for Air) wrote:

'This was the worst instance I know of the scandalous practice of sending Active Service pilots who were certainly not fit for their jobs. The Squadron flew to France, and I think that of the twelve machines that left Gosport only two or three arrived safely. The remainder were scattered about England and France in various stages of damage both as regards pilots and aircraft. It was some weeks before this unit could be reorganised, while the cost to the Exchequer of this more haste less speed policy must have run into many thousands of pounds. It would have been far better to have deprived the needs of France for a few weeks longer, and finally sent a unit which could have been of immediate use.'

Not that Balfour was a paragon of virtue, as he was the first to admit when he invited his first passenger up in a Morane from Gosport. Monoplane Moranes were tricky beasts but more importantly they were built for one occupant – a passenger could only be carried by sitting behind, but on the same seat as the pilot, with arms clasped around his waist.

'We had the whole length of the big aerodrome to get off in. Directly I had got the machine taxying along the ground trying to gather speed for take-off, I knew that I was in for a rough time and had bitten off more than I could chew. The Morane with two up was a very different proposition to solo. I had sufficient sense once we were off not to try and climb at anything other than a very gradual rate, and not to endeavour to turn. Eventually we staggered half-way over the Solent, reaching 1,200ft. I slithered slowly round back to the aerodrome, only too anxious to get once more on the ground. When we were over the shed I shouted to my passenger to look out as I was about to switch off and glide in to land. I cut the Gnome and the next moment the Morane dropped its nose almost vertically. We hurtled down towards earth, and I knew that if I flattened out too soon we were helpless, for the Morane had a habit of viciously dropping its wing if glided too slowly, and this would be accentuated with the heavy load, while if I did not flatten out soon enough we would go straight into the earth. At what seemed like the crucial moment I pulled the stick back. Unfortunately I was about two feet high. There was nothing to be done except to await the inevitable. We hit with the right wheel and wing tip, and did a complete half-loop on to the ground finishing hanging upside down in our belts with petrol running all over us. Fortunately no fire broke out, and we extricated ourselves without damage. When Smith-Barry, quite rightly, told me not to try to show off in front of people, I knew that his rebuke was thoroughly deserved.'

No flying organisation should ever be run on such fixed or blinkered lines that it is unwilling to consider new ideas from unorthodox minds. After four years of teaching men to fly, and losing 15,000 aircraft in the process, the RFC had to wait on Smith-Barry to come up with the stunningly simple idea that a school should be established to instruct instructors. Until then, as he noted in forthright manner:

'. . . no attention whatsoever was paid to the fundamental importance of instruction in the mere manual part of flying. This was left to those who were resting, those who were preparing to go overseas, and those who have shown themselves useless for anything else. The first two classes had other interests paramount; the third had no interests at all . . . No Scheme was laid down for them and no instructions were given them, and they therefore looked at their work as the merest drudgery. It is submitted that a very good way to remedy this . . . would be to have a school for training instructors where they could (a) have their flying brought up to the very high standard necessary before they can teach with confidence and ease, and be combed out if they do not speedily reach this standard, and (b), be given definite lines upon which to instruct.'

The new ideas Smith-Barry implemented at Gosport have stood the test of time. Firstly, he got rid of the multiplicity of elderly aircraft then handed down to serve as trainers, building up his School around a standard fleet of dual-control Avros built for the purpose. Smith-Barry emphasised the vital importance of dual controls so that pupils could take charge of their aircraft gradually and be shown that steep manoeuvres and even spins were not fatal. Then he introduced the simple but effective 'Gosport tube' between back and front cockpits so that the instructor could explain what was happening rather than having to rely on hand waving and shouting during the few seconds of quiet before a stall. Thus endowed, Gosport instructors taught skills previously never covered such as spinning and aerobatics, with the result that their students passed out far better prepared for immediate transfer to the frontline. On the other hand, Smith-Barry's Flight Commanders with dual control machines were now held personally responsible for all crashes that took place in their Flights.

Once Gosport was converted into the Instructor's School for the RFC, Smith-Barry produced a booklet entitled 'Notes on Teaching Flying' to spread his gospel. There had been written instructions before but they had been somewhat limited as illustrated by the following gems issued with the 1911 Curtiss aircraft:

'When the mechanism is facing into the wind, the aeronaut should open the control valve of the motor to its fullest extent, at the same time pulling the control pole toward his middle anatomy. When sufficient speed has been attained, the device will leave the ground and assume the position of aeronautical ascent.

'Should the aeronaut decide to return to *terra firma*, he should close the control valve of the motor. This will cause the apparatus to assume what is known as the "gliding position", except in the cases of those flying machines which are inherently unstable. These latter will assume the position known as "involuntary spin" and will return to earth without further action on the part of the aeronaut.'

Smith-Barry had more practical advice in mind because he wanted to get away from the old training regime when, in the words of one pilot, 'Often you had an instruction manual shoved in one hand and a joystick in the other'. After reiterating the necessity for dual control, his 'Notes' stated many good points such as 'the next and most important thing is that quite half the dual control is administered after the pupil has gone off alone . . . as he will not appreciate the details that are shown him. In this way bad habits are corrected before they have time to get fixed.'

Smith-Barry's supreme object was to teach that 'most of the supposed dangers are not dangerous at all when properly tackled', and that 'it would seem a simple matter for the pupil to be taught, chiefly by example, to be frightened of nothing connected with flying on this side of the lines'. Smith-Barry's greatest legacy was to teach the RFC, and through it the rest of the world, that in knowledge lies safety.

The RFC and Royal Naval Air Service were amalgamated to form an independent RAF in 1918. Its founding father, Sir Hugh Trenchard, prepared a White Paper on the future of the new service including key paragraphs on the 'Extreme Importance of Training' which concluded that 'we must use every endeavour to eliminate flying accidents, both during training and subsequently'. Trenchard had good cause to emphasise this latter point because, despite Smith-Barry's efforts, the RAF's first year was one of its worst for crashes. Furthermore, as aircraft increased in size, so the death toll rose when they did have accidents. Thus on a test flight over Maxstoke in August 1918, fabric was seen to stream back from the lower wing of a Handley Page 0/400 bomber which then nose-dived to earth killing two officers and five air mechanics. However, such technical failures only accounted for around a fifth of the 324 accidents that plagued the RAF in the single month of August 1918. Pilot error and collisions were responsible for 172 accidents, as when a 0/400 turned over on its nose after colliding with an Avro on the ground. To add further insult to professional pride, though the giant bomber was still sitting there the next day with its tail in the air, a Spad taking off crashed into the wreckage with sufficient force that the 0/400 was knocked on to its back.

One RAF summary was quite scathing about the fact that 'accidents during the last three months of 1917 were sufficient to buy new gaiters and spurs for each and every pilot and observer in the service!' Yet no matter how the cost was calculated the RFC, and later RAF, was paying the price that all air forces have to meet when they try to expand rapidly to meet the needs of war. In striving to turn out thousands of pilots every year, both to create new units and

Above:

Not much in the way of personal protection – an aircraft from 23rd Training Wing at Scampton, Lincs, having bitten the dust on 14 March 1918.

to keep existing ones up to strength, corners were cut and lessons not learned as thoroughly as they ought.

Once World War 1 ended and the military found that it had more training time available, the flying accident rate should have reduced but it did not. In 1921, when the US Army Air Corps began keeping track of mishap rates, it had 361 major accidents during a total of 77,000 flying hours. This equated to 46.7 accidents per 10,000 flying hours and, if the USAF flew like that today, it would be crashing 1,350 aircraft a month and would use up its entire inventory in seven months.

The RAF was equally concerned about its accident losses. Between 1921 and 1925 the RAF suffered 258 serious accidents which, allowing for differing definitions of what constituted a 'serious accident', equated to 9.1 per 10,000 flying hours. Another 57 serious accidents occurred in the RAF during 1926, which prompted Maj C. C. Turner to write an article in *The Daily Telegraph* on 7 September chiding the Air Ministry for not compiling and publicising 'the completest possible statistical data showing the type of machine crashed, dates, times and weather conditions, the causes assigned to the accidents, the nature of the flights and the status of the pilots'. The worst months for accidents every year were August and September, which might indicate that everyone was short of practice after returning from their summer holidays, but Turner felt that no improvements could be made until someone looked at the overall picture and identified the groups of pilots or particular flying profiles most at risk.

Between 1921-26, 40 fatal accidents had occurred during RAF flying training which Turner felt might have been due to elderly training machines. However, while acknowledging that 'in the absence of complete information it is unwise to make sweeping judgements', Turner felt that the RAF could be pushing its new pilots too far too early.

'There is in some quarters a feeling that in the RAF the same spirit of daring is inculcated as was necessary in intensive training during the war. The young pilot is expected to take naturally to the air like a swallow, and the slightest suggestion of inquiry or nervousness on his part is roughly regressed. It cannot be too strongly insisted that the acquisition of the highest piloting skill calls at all stages for deliberation, and that everything in the nature of derring-do is as inimical to the attainment of excellence as it would be in the Staff College. Of course the RAF must have daring officers . . . but there must be something added to dash and gallantry. In wartime let it be admitted, the need for a large stream of reserve pilots calls for intensive methods, to which much must be sacrificed. We are enjoying peace, however, and there is no need for intensive methods in training to fly. This is not the time for recklessness. Further, the depressive effect on the Service of unnecessary accidents must not be forgotten: it is inimical to true efficiency.'

The files show that Sir Hugh Trenchard instituted a subsequent study into the causes of flying accidents because, as he wrote to *The Times* in February 1931, 'there is no subject which caused me greater anxiety during the decade in which I held the post of Chief of the Air Staff and none to which I devoted more time and thought'. The main causes of accidents identified by Trenchard's staff in 1927 will bring wry smiles to the faces of military pilots today because they have not changed much in 60 years – inadequate supervision of flying, too much interference with flying training, shortage of senior Flying Officers on Flights to guide new arrivals, too much clerical work to detract from the primary task of flying, inadequate allowance for station secondary duties and too many breaks in the early part of a pilot's flying career. But between 1924-26, half or more accidents every year were caused by a pilot's 'error of judgement', and even more disturbing a further 10% of accidents resulted from 'contravening orders'.

A cross-section of some of the RAF's 22 fatal accidents between 1 January and 12 May 1931 illustrates the point:

- No 84 Squadron Wapiti – attempted experimental landing by light of flares – flew into the ground.
- No 2 Flying Training School (FTS) Vimy – lost his way, landed in dyke in mist and darkness.
- No 204 Squadron Iris – second pilot not qualified on flying boats. He refused to hand over control to first pilot before landing which he then misjudged.
- No 33 Squadron Hart – flew into flag pole.
- No 2 FTS Avro – while pilot practising 'crazy' flying at about 200ft, aircraft stalled and hit the ground.
- No 29 Squadron Siskin – disobeyed orders on leaving sector; carried out reckless low level aerobatics and hit the ground in low turn.
- No 602 Squadron Wapiti and Avro N – practised unauthorised mock fighting.

In his letter to *The Times*, which followed a public outcry over the 64 fatal RAF accidents the previous year, Trenchard made a plea that the number of accidents should be offset against the increased amount of flying then being undertaken and that the nation should look at accident trends over a period of years rather than concentrate on the odd bad batch. Unfortunately the great man was on dicey ground here because, while the very bad patches of eight fatals in 12 days were rare, the Service had lost 76 men in 1928, 42 in 1929, 64 in 1930 and would lose another 74 in 1931. If the 'good patch' was only 42 dead, then something was wrong.

Trenchard had an intimation of where the trouble lay. 'Much is being made,' he wrote, 'of the recent accident to a large flying boat at Cattewater . . . If my information is correct, that was due to a direct contravention of orders on the part of an individual officer. I would say outright that no precautions and regulations can preclude the possibility of disaster if orders are disobeyed. One can only see to it that it is brought home to all concerned that they expose themselves, but also others, to the gravest risks if they do not strictly comply with the orders given them in the interests of safety.'

The trouble is that orders are not the be-all and end-all of flight safety. If life were that simple, the military could simply repeat the order put out by the US Army's Chief Aviator in 1934: 'There will be no more accidents!' Such an injunction might give a great deal of satisfaction to the man who sends it, but it took one of Gen Westover's zone commanders to put things in a proper perspective when he signalled back: 'There will be no more flying'.

RAF commanders in the 1920s and 1930s, who had earned their spurs over the trenches, probably went too far the other way in promoting the 'can do' spirit. For example, it was a great time for air displays to show off the RAF, and at one such Andover show in 1926, Flg Off John Boothman in a Snipe put on a unique programme of combined aerobatics with another CFS flying instructor, Flg Off D'Arcy Greig. As the climax, Greig's Cirrus Moth began a spectacular falling leaf from 1,000ft while Boothman performed an inverted falling leaf from 2,000ft immediately above him. Just as he began rolling out at a safe height, Boothman was horrified to see the Moth dissolve into a cloud of sticks and canvas as it hit the ground. There was an awful moment of silence and then from the dark depths of the wreckage came a single very clear and very rude word. One high ranking witness was reported to have been in favour of immediate disciplinary proceedings against Greig, a man well known for such stunts as wing-walking in the air while reading a newspaper. But the commandant of CFS shrieked with laughter and offered to find another aircraft if Greig would do it again.

Five years later, a young 20-year-old Douglas Bader took on a dare to 'beat up' Woodley aerodrome near Reading. He crashed his Bristol Bulldog in the process and his injuries led to the amputation of both legs. The laconic note in his logbook was a classic flight safety one-liner: 'Crashed slow-rolling near ground. Bad show.'

Such episodes are the bed-rock of RAF tradition and it is a dull soul who is not enlivened by tales of bombing other units with toilet rolls, flying low through hangars or formating a few feet from the cabs of express trains. But in setting such an example, natural pilots bear some responsibility if lesser mortals feel beholden to follow suit and fail to get away with it.

36

There was a wise World War 1 maxim: 'Keep their tails up on the ground, and they'll look after themselves in the air'. But it is one thing to try and overcome the fear of combat; it is another altogether to exceed the limits for their own sake. Robert Smith-Barry certainly had no time for taking gratuitous risks in the air — he would have court martialled a pilot for doing aerobatics in a Sopwith Pup at 200ft over Piccadilly had not the offender been immediately posted to France. It is true that his 'Notes on Teaching Flying' stated that advanced pupils should be allowed 'to fly exactly as they chose, their experiments being limited only by the state of their nerves'. Yet to be fair to Smith-Barry, he only believed that 'nothing that a pilot may do in the air is dangerous' provided '*he knows what he is doing and what the result will be*'. This is the crux of the matter, and in the final analysis flight safety relies on an attitude of mind rather than mere adherence to orders. If military crews fly about serenely and sedately, they will probably be unable to cope with flying to the limits against an enemy. On the other hand, if too many people and aircraft are written off in the peacetime pursuit of 'punchiness', there might not be enough of either left to fight a war if it comes. Finding the proper balance between derring-do and derring-don't is where an air force commander earns his money.

Having said all that, safety improvements were certainly made in the interwar years such as ensuring that crews instinctively carried out drills in the event of fire in the air. But when World War 2 came, it was back to the old days of 'sausage machine' training. The Americans lost more aircraft and crews in training and routine flights than in combat. Their worst year was 1943 when they had 20,399 major mishaps in the US alone, killing over 5,600 aircrew. That same year, the Inspector General of the RAF, ACM Sir Edgar Ludlow-Hewitt, wrote that 'our incidence of aircraft accidents is the enemy's best friend and ally in the promotion of his air war against us, and probably accounts for about as many of our aircraft as the *Luftwaffe* itself'. Throughout the RAF as a whole, between 1939 and 1945 a total of 80,000 aircraft were accidentally destroyed and a further 160,000 were involved in minor contretemps. As only some 125,000 aircraft were built, many must have gone through the accident mill more than once. But much worse than the loss of aircraft, which could be replaced or rebuilt, was the loss of life. Between 1940 and 1945, the home-based RAF lost 25,271 aircrew in non-operational flying accidents, and Ludlow-Hewitt estimated that the total loss, including overseas RAF elements, was 'in the order of 10,000 trained and semi-trained flying men each year. Here is a foe within our gates.'

The RAF's peak rate of 27.13 major accidents per 10,000 flying hours occurred in 1941 but even though the rate was nearly halved by 1945, during the last year of the war RAF Transport Command suffered 1,531 accidents of which 738 were positively due to some form of human error or indiscipline either in the air or on the ground. Fortunately these graphic examples of how operational potential can be eroded by accident wastage were not confined to the Allied side; the Germans lost 25% of their total aircraft production in the last 12 months of the war from ferry flying accidents alone.

As early as 1923, an Inspectorate of Accidents had been set up within the British Air Ministry's Department of Civil Aviation to report on all accidents to both service and civil aircraft. In 1937, coincident with the expansion of British military flying, this branch was expanded to encompass seven inspectors of accidents working under the Chief Inspector, Wg Cdr V. S. Brown. His staff and workload increased considerably during the war such that by 1944 Vernon Brown was still Chief Inspector but now in the rank of Air Cdre. However, it took postwar developments such as the jet engine and supersonic flight to shift the emphasis on both sides of the Atlantic from reacting to mishaps to preventing them from happening in the first place. The increased efforts to find out where problems might lie and to educate people accordingly paid dividends. The RAF's major accident rate was about 8.5 per 10,000 flying hours in 1947, it was 4.1 in 1957 just after the Directorate of Flight Safety was established, and thereafter it fell to 0.68 in 1967, 0.34 in 1977 and 0.28 in 1987 a considerable improvement.

Yet despite the noticeable improvement, the Chief of Air Staff was moved to remark in 1966 that annual accident wastage was still equivalent to the loss of two fighter squadrons. Since 1918 the official means of investigating accidents to RAF aircraft has been first the Court of, and then the Board of Inquiry. Board rules and procedure have changed little since 1924 and are now based in the Air Force Act of 1955. Generally the parent command of the aircraft concerned convenes a Board if there has been a fatality, if the cause has not been established beyond reasonable doubt, or where negligence or default is suspected.

The Board acts as a fact-finding tribunal but it does more than simply determine accident causes. In a disciplined service, the performance and actions of individuals must also be judged in a firm but fair fashion. The Board of Inquiry therefore serves as a convenient, all-round tool from whose findings a commander should be able to judge all operational, administrative, technical and disciplinary facets of an accident.

Following an accident that is serious enough to merit a Board of Inquiry, the convening authority details an officer to act as president. The president is normally drawn from a current aircrew background, at least a wing commander in the case of a fatal accident, and not junior in rank or seniority to any officer whose conduct, character or professional reputation may be called into question. The president will be assisted on the Board by two or more officer members junior to him — one will usually be aircrew and the other an engineer. They should all have recent experience of the aircraft type or role under investigation, and ideally they should come from another station so that their judgement is not clouded by personal factors.

No two Boards are ever the same, but in general terms an RAF accident is investigated along the following lines. Once assembled, the Board begins what can best be termed as a gathering of facts and clearing of thoughts phase. This culminates in the dispatch of an Interim Report by signal within 96hr (or 48hr if possible) to give a preliminary assessment of the cause if known, comment on the integrity of the aircraft and its systems, and assess the validity of operating procedures. Occasionally a major structural defect can come to light — as when metal fatigue caused the starboard wing to fall off a Buccaneer while it was flying over Nevada in February 1980. The Interim Report can alert higher authority to the possibility of grounding an entire aircraft type until the full facts are known.

As early as possible after assembly the Board considers the need for detailed examination of both crash scene and wreckage. In this, as in many other endeavours, it is greatly assisted by outside specialists such as those from the Air Accidents Investigation Branch at Farnborough. As a further example, requests for assistance from the Institute of Medicine's Aviation Pathologist or Behavioural Scientist are compulsory when an accident involves fatalities or suspected human error respectively. In sum, the Board is never on its own.

The next stage is to take evidence from witnesses. Evidence is given under oath but a Board of Inquiry, not being a court of law, may receive any evidence it considers relevant so long as it is the 'best' evidence available. The important point is that evidence given to a Board is privileged and therefore will not be made public. This is an aspect the RAF guards jealously because it encourages honest reporting.

Having sifted all the evidence, the Board is ready to record its findings in a form which aims to provide the reader with a logical outline of what happened together with all supporting evidence. Board proceedings start with a narrative of events followed by a diagnosis of all possible causes. The Board assesses the relative importance of each cause, and accepts or discounts it on the basis of the evidence. For example, a sudden loss of engine power might have resulted from contaminated fuel, so the fuel analysis results must be obtained. This diagnostic process of elimination continues until the Board is in a position to make a statement along the lines of 'the primary cause of the accident was the pilot's failure adequately to maintain airspeed during the latter stages of the approach. A contributory cause was the presence of moderate turbulence'.

The Board then considers other factors such as degree of injury, whether service personnel involved were on duty at the time, compliance with orders and instructions, the effectiveness of aircraft escape facilities, damage to property, and consideration of human failings. Finally, the Board makes its recommendations with the aim of preventing a repetition. The Board then presents its findings to the Station Commander for his comments, before the whole package is progressed through the command chain until the Board is finally signed off by the Air Officer Commanding-in-Chief. At any stage in this process, the Board can be re-convened if errors or omissions are found which need to be resolved.

This then is the broad framework of the RAF's accident investigation procedure. It relies on current aircrew and engineers, supported by a wealth of specialist talent, to investigate the circumstances of an accident, to determine the cause or causes, and to make recommendations on remedial action to prevent a recurrence. This final factor is most important. A Board is never regarded as a 'hanging party' or a purely reactive process designed simply to close the stable door. Rather its members are a team who fully understand what the crew was trying to do, are cognisant of the pressures they were under, and who are mindful of the need to find out what went wrong so that the lessons can be disseminated to others throughout the Service as expeditiously as possible.

Suffice to say that despite the advent of more reliable aircraft and 'all-singing, all-dancing' avionics over the years, the RAF is unlikely ever to eliminate flying accidents completely. As long as there are human beings in cockpits, there will always be the chance of human error or pressures that overstretch a human crew in certain circumstances. Tragically, there also seem to be a very few who insist on pushing their luck too far.

Not too long ago, two Lightning interceptors were programmed to pre-position at a civil airport for a flying display the next day. During the morning, the authorised display leader of the pair received an unofficial telephone request to overfly a recruiting display being held at a nearby seaside resort en route to the civil airfield. As it seemed a good opportunity to show the flag, the pilot discussed the suggestion with his authorising officer but the latter was adamant that, at such a late stage and without official approval, the informal flypast could not take place.

Came the planned take-off time and the second Lightning was still being prepared, so the display leader was charged to go ahead as a singleton. Some 10min after getting airborne, his Lightning was seen to fly past the recruiting exhibition at between 100-200ft. The pilot carried out a wingover at the north end of the bay, and then descended to make a second pass from north to south in the landing configuration. In the latter stages of the pass he engaged reheat, raised undercarriage and flap, and made a left-hand climbing pull-up to a height of about 1,500ft heading out to sea. He then reversed the turn and descended, heading back towards the headland and the southern end of the bay. This headland had almost sheer rock faces rising to 255ft. Witnesses saw the interceptor make a tight descending right turn at a high angle of attack to narrowly miss the cliff face. The Lightning crossed the coastline again heading out into the bay at very low level, and as the pilot apparently tried to roll the wings level, the aircraft pitched up, appeared to 'hang in mid-air', and then the nose dropped. The aircraft yawed and rolled to the right as it descended, hitting the water and killing the pilot instantaneously.

As examination of the wreckage found nothing to be wrong with the aircraft, it must be concluded that the pilot stalled in his attempt to keep clear of the cliffs. In succumbing to the temptation to give a really 'punchy' unauthorised display, he paid with his life for flying far lower than even an approved display warranted over terrain whose natural hazards had not been properly considered beforehand. What made an otherwise experienced and respected display pilot and leader suddenly kick over the traces when a moment's thought would have told him that even if he had survived, he was laying himself open to disciplinary action? If you can answer that, you can name your price to any Air Board in the world. But until we reach the stage where we can understand and predict the intimate workings of every human mind, we can only recall the old adage: 'There are old pilots and there are bold pilots, but there are no old, bold pilots'.

Finger Trouble

Two months after the start of World War 2, a chastened flying officer sat down at RAF Harwell to compose the following letter:

'15 Nov 39
To: OC No 148 Squadron
Subject: Report on Collision with Balloon Cable.
1. At about 20.20hrs on 13 November 1939, I was flying an Anson aircraft at about 1,500ft. Owing to navigational errors I did not know my exact position.
2. I was flying through small patches of cloud, mainly by my instruments, when suddenly I felt a bump and was thrown forward against the stick and windscreen. The aircraft banked to the left and put its nose down in a spiral. I then saw lights below and applied opposite rudder quickly, at the same time easing the stick back slowly until the aircraft was flying level at about 1,000ft. After feeling my lateral control carefully, I proceeded towards base. At the time of the collision I was flying at about 120mph. I did not see the balloon or balloon cable either before or after the accident. I am still not certain of the position but from subsequent homing bearings I presume it to have been within the London balloon barrage.'

With part of his port wing sliced away, it was impossible for the unfortunate pilot to hide his transgression, but doubtless he and other members of his unit learned much from the incident. Unfortunately there must have been many other aircrew who, having been pushed through the wartime training schools

Below:
Legacy of a collision between the wing of a No 148 Squadron Anson and a barrage balloon cable on 13 November 1939.

and thereafter thrown into the operational deep-end with the barest minimum of trial runs first, never had time to learn for themselves or from the experience of others. It was not just survival of the fittest — it was also survival of the luckiest.

The promotion of flight safety awareness stands or falls by the effectiveness of its publicity, and the most effective publicist in the cause of RAF aircraft safety in World War 2 was Plt Off Percy Prune. This cartoon figure's creator was Bill Hooper, who in 1941 was serving with No 54 Squadron at Catterick where the squadron was catching its breath after the rigours of the Battle of Britain. On the strength of his illustrations in a book of hints and tips for fighter pilots, Hooper was asked by Flt Lt A. A. Willis (otherwise known as the playwright, Anthony Armstrong) at the Air Ministry to evolve a dim-witted character who would carry the flight safety message to potential 'Prunes' in an official training manual to be called *Tee Emm*.

Thereafter, Plt Off Prune became as famous within the RAF as any of the 'aces'. He was the fool and permanently bone-headed clot who invariably made a mess of everything he set out to do, yet despite every adversity, Prune remained fatuously exuberant as he sallied forth to make another crass blunder. In the words of his creator: 'From the smoking wreckage of his latest "prunery", miraculously as safe as ever, (Prune) contemplated with a hurt and puzzled detachment the incomprehensible eccentricities of a world which was never quite within his grasp; a world in which undercarriages never came down of their own accord before his landings, a world in which his famous finger went unerringly to the wrong "tit" — and he never understood why. Prune tried but never learned. He was willing, but wet. He was dutiful but dumb.'

Prune never progressed beyond the lowly rank of pilot officer but he served till the end of the war as an awful warning to others. His escapades were risible and while Prune's casual approach to operational flying was often an echo of real pilots' errors, Hooper's genius lay in the outrageous excuses he made Prune give for his sins: they had just enough grains of truth in them to trouble more than a few frontline consciences. Prune defined a good landing as 'one you can walk away from' and a perfect three-point landing as 'two wheels and a nose'. Aided and abetted by an equally motley crew, Prune could reduce a Board of Inquiry to despair. For example his navigator, Flg Off Fixe, was heard to say after returning with the entire tail unit shot away that they flew back so low over the Channel that 'we had to stand up to see over the waves'. On another occasion his wireless operator, Sgt Backtune, told a Board. 'Well, you see Sir, nobody knew what actually hit us because we were all eating our sandwiches at the time'.

As Prune's fame grew, he became the main medium for passing on flight safety messages provided by experts from within the RAF. Thus when fire in the air became a particular concern, it was felt that a serious article on the subject would not get across as widely as it ought. So Prune was made to boast in the Officers' Mess that the official guidance on dealing with fires in the air was rubbish and that in his experience, the only good way to extinguish a fire was to lower the nose, descend at maximum speed, and blow the flames out!

When Bill Hooper met a Free French pilot, 'said to be the "unluckiest" to have flown under the Cross of Lorraine', Hooper was staggered to hear that the

Frenchman had just 'got an 'Un'. Upon checking with the French Commandant he was told: 'Yes Bill, but 'e didn't shoot 'im down, he *collided* with 'im over the Channel'. This gem gave Hooper the idea for a Gallic Prune and so 'Aspirant Praline' was born.

But Prune's most notable achievement was to become Patron of the Most Highly Derogatory Order of the Irremovable Finger, an accolade awarded monthly by *Tee Emm* for any instance of outstanding ability to do the wrong thing at the wrong time while flying. From then on, the term 'finger' became a universally understood pithy admonition to get your act together. There were certainly many examples of 'finger trouble' within the RAF. At the most extreme were those accidents caused by the fog of war, such as when two Beaufighters intercepted a Stirling bomber in 1941; one of them shot it down and then the second Beaufighter shot down the first. But outside the combat environment there was much less excuse for sins of omission. On 11 October 1944, a flight lieutenant with a DFC to his name took off from Nuneaton airfield on the first production test flight of a Spitfire Mk IX which had just been rolled out of the Castle Bromwich aeroplane factory. As the weather was very bad over most of the Midlands, he had to climb above 12,000ft to get above cloud. Shortly after take-off, the Spitfire was heard to 'open up' and it was assumed that the pilot was going into a power dive. The Spitfire then appeared out of the cloud in an almost vertical dive, crashing into the middle of Nuneaton airfield. The airframe and engine were smashed into small fragments and the pilot's remains were found intermingled with the wreckage.

Examination of the shattered Spitfire failed to reveal any evidence of technical or structural failure which would have accounted for the accident. Subsequently numerous people declared that the pilot had been suffering from some kind of illness and that on the day of the accident he had told ground staff that he was 'not at all well'. It was also discovered that Castle Bromwich test pilots shunned the use of oxygen masks during ordinary production test flights up to 20,000ft, preferring to put the aircraft oxygen pipe into their mouths only when they felt the need.

The Chief Inspector (Accidents) therefore concluded that the circumstances of the accident 'strongly suggest loss of control due to the incapacity of the pilot. This incapacity was probably caused by oxygen lack coupled with a sub-normal physical condition'. Such bodily 'finger trouble' proved that it was just as important to know and respect your own physical limits as those of your aircraft.

During World War 2, some aircraft had a better accident record than others: for example, the Stirling's accident rate per hours flown greatly exceeded that of the Lancaster and Halifax. It was easy to blame the aircraft itself for this, particularly because the Stirling had a tendency to swing during take-off and landing and its relatively weak undercarriage could not withstand heavy strain. However, more detailed examination of the statistics showed that there was a human error dimension in many mishaps. In 1942 and the first four months of 1943, there was almost a glut of Stirling accidents during which the pilot landed with the tailwheel retracted, the flight engineer either having forgotten to lower it or failed to check that it had locked down. Furthermore, although word got about that the Stirling was more prone to fire in the air, it was only more prone to the fire spreading because crews mishandled their switching drills. Air locks due to incorrect operation of the petrol cocks also led to several disastrous accidents in 1943, and to add insult to injury, most of these occurred on operational squadrons rather than conversion units, highlighting the need for constant updating of knowledge on aircraft systems.

'Finger trouble' was often the fatal link in a chain of events leading up to an accident or series of accidents. On 26 January 1944, Stirling EH933 of 1660 Heavy Conversion Unit (HCU) took-off from Swinderby on searchlight/fighter duties combined with a navigation exercise, the whole to be flown at 18-19,000ft altitude. About 2¼hrs after take-off EH933 was heard in the vicinity of

Below:
The dramatic effects of structural failure in the air. Halifax LW382 took-off from Linton, Yorks, in the early hours of 6 June 1944 as part of the D-Day operation. For some reason that the Board of Inquiry could never determine, the Halifax went into a steep dive as it passed over Norfolk. The evidence showed that the pilot pulled 6g in an attempt to recover, but LW382 could not stand the strain and it broke up. The crew never had a chance to bale out and as the aircraft crashed, its bomb load exploded.

Winsford between Sidmouth and Ilfracombe heading west. It was on the fourth leg of the exercise but it was then heard to turn southeast with the engines spluttering. As the Stirling lost height, the engine noise stopped and 4sec later — it was just after midnight — the aircraft crashed and burst into flames at Bridgetown in Devon.

An Accidents Branch Inspector found the main wreckage lying inverted with a trail of disintegrated debris extending back for some 3½ miles. The aircraft appeared to have broken up in the air, and Farnborough was able to deduce that the rear fuselage had broken away at about 14,000ft under loading conditions indicative of a dive. But why had the aircraft entered such a steep dive that an almost brand new bomber had broken apart? The experts could only postulate that the aircraft had been in thick cloud for most of its flight, the resulting turbulence made for difficult flying, and that the pilot, who had only 250 total hours in his logbook and had only made one cross-country flight in a Stirling before, lost control to such an extent that the ensuing high speed dive resulted in massive structural failure.

Four months later, two more Stirlings crashed in similar circumstances. Both were on training exercises and the second, LK517, belonged to 1654 Heavy Conversion Unit at Wigsley. On 31 May the crew was briefed to fly LK517 for 5hrs on navigational and bombing exercises, and they were warned by the Met man that cumulo-nimbus clouds with tops up to 20,000ft were forming over the Pennines and could be expected to the east. At around 16.45hrs, the Garrison Engineer at a PoW camp near Darlington heard an aircraft and from the unusual noises made, it appeared to be in difficulties. Immediately afterwards he heard two muffled explosions — 'as soon as the explosions occurred, the engines of the aircraft opened up and the noise produced was such as I had never heard from any aircraft before'. He then saw pieces of bomber falling out of the clouds: all that remained — the fuselage forward of the entrance door, the mainplanes less tips, and the four engines — came down between Middridge and Shildon, Co Durham. All seven crew members died.

The accident investigators now found themselves with three Stirlings that had disintegrated in remarkably similar fashion, but there was no evidence to give a firm idea of which part of the structure failed first or why. All the Board inquiring into the loss of LK517 could surmise was that 'loss of control when flying in cloud in which icing and extremely bumpy conditions were to be expected . . . may have resulted in a steep dive at very high speed in which overspeeding of the motors occurred. A violent pull out of the dive, probably on catching sight of the ground, may have resulted in an initial structural failure followed by the disintegration of other portions of the airframe.'

There the matter rested until Stirling LK499 of 1653 HCU was carrying out a night training exercise on 12 September. Flown by a pilot with only a total of 18hr 35min on type, the aircraft left Chedburgh in Suffolk at 21.15hrs on a navigation exercise to Rugby, Bristol and thence to Plymouth. All seemed to go according to plan until the Stirling was heard just north of Plymouth flying at approximately 5,000ft. Shortly afterwards, a flash on the ground was observed in the direction of Leemoor, Devon, which proved to be LK499 crashing following structural failure in the air. A trail of airframe pieces extended some 900yd to the west of the main wreckage.

The weather on this occasion had been good and although the pilot's flying ability had not been assessed highly, there was no apparent reason why he should have let a momentary lapse of control develop into a very high speed dive. The only new evidence found by the investigators was that the rear fuselage had become detached from the rest of the aeroplane in download.

Then on 19 October, LK207 took-off from Tempsford at 10.59hrs on a local air test. Ten minutes after getting airborne, witnesses heard an unusual noise followed by a sight of the aircraft spiralling down with pieces detaching. The Stirling crashed at Potton, Bedfordshire, killing all occupants. As with EH933, the main wreckage was found lying inverted having struck the ground in a flat attitude with little forward speed. Detached portions, including both outer wings, rear fuselage, tailplanes and elevators, were strewn up to two miles away.

Once again the wing detachments had occurred during download but for the first time there was now enough relevant wreckage to show that the fuselage flooring had collapsed in compression, which also indicated a downwards failure. In other words, the break-up of the Stirlings could no longer be attributed to excessive forces imposed by pilots as they strove to pull out before hitting the ground. Rather it was very opposite — for so much to have collapsed or broken off in download, the aircraft had somehow or other 'bunted' or got on to its back.

Flight tests were then made which showed that, if the Stirling was dived beyond its limiting speed, a nose-heavy tendency progressively built up. But why did a variety of pilots, including the one in LK207 who had accumulated 382 night flying hours on Stirlings and Halifaxes, lose control so drastically? The answer came from Air Cdre Vernon Brown, the Chief Inspector (Accidents): 'One of the reasons why so many pilots have lost control is, without the slightest doubt, the result of failure to wear their harnesses'. The Stirling's seat harness was so restricting that pilots got into the habit of releasing it after take-off and strapping in again only for landing. In between they were not only more comfortable but also they were free to move, especially forward. Thus, if a pilot lost control and his Stirling went into a dive, he would have fallen forward on to the control column. This would have pushed the Stirling into an even steeper dive and eventually it would have bunted over with the pilot helpless to regain his seat, leading inexorably to the collapse of a stout aircraft.

'I am quite tired', continued Vernon Brown, 'of asking pilots if they wear their harnesses and receiving a steady look and solemn reply that, "the regulations are quite clear upon the subject, sir!" ' We cannot know how many of the 641 Stirlings lost in World War 2 never came back for reasons as silly as pilots not wearing their harness in the air but it was probably far too many. Bomber Command should not have had to await the introduction into service of the more comfortable 'Q' type harness before such stupid accidents were eradicated.

Right:
Short Stirling of the Conversion Flight at Oakington, Cambs, in 1942. This aircraft and all the crew on board were to perish soon afterwards when the Stirling was flown into some high ground on a night cross-country exercise.

'Finger trouble' is basically lack of awareness. Lack of awareness is not just confined to the ignorant — it can also come about when experience slips into overconfidence.

For the second half of World War 2, the elite component of Bomber Command was the Pathfinders. Only the ablest and the best were detailed to point the way for the Main Force, and in 1943 one of their number led his squadron off from Oakington to attack the Ruhr.

'We had instilled in all of us a very strong sense of timing, which had to be to the second. So as a rule, I would always get out of dispersal in plenty of time. I would get very testy if things weren't exactly right come the take-off time.

'On this occasion something had delayed me and I was late. I was in command of a heavily laden Stirling which was renowned for the fact that on a cold night the engines took a long time to warm up. The orders were very strict that one could not move off until the oil temperature was above the minimum, but I was quite determined to make up the time. I got to the end of the runway first, applied my brakes, and did my run-up and magneto checks whilst waiting impatiently for the oil temperature to rise to the required degree.

'The flight engineer complained bitterly as I was making all the preparations for take-off, but I ignored him. The navigator, was giving me the countdown; as soon as the needles on the temperature gauge started moving I would be off. I was very proud of the reputation I had earned for exactness of timing. Take-off time arrived and the oil temperatures were still somewhat below the criteria, but I opened the throttles and very shortly the tail was up and I was heading down the short runway into a brisk headwind. The wheels lifted off, up came the undercarriage and, as it was the short runway, I left the flap down a little longer than usual. To my absolute horror I found that all my controls were

completely frozen. I was quite certain I had seen the groundcrew remove all the external locks before I had climbed on board.

'We were climbing away at a very gentle angle and I could not understand why I could not move the controls. Fortunately the aircraft maintained a perfect climbing attitude while I tried to sort it out. I reached for the elevator trim wheel; one glance told me it was in the normal position. We were still climbing at full power and a touch of nose up trim enabled me to increase the rate of climb. As my hand came away from the trim wheel I touched the autopilot control and, on looking down, I found to my horror that it was "Engaged"; I had taken off with "George" in control.

'Being one of the most experienced pilots in Bomber Command, and having done an instructional tour, meant nothing. In my hurry and in the conceit that is deep within many of us pilots, with our belief in our own infallibility, I hadn't done my pre-take off checks. I was able to climb to about 1,200ft at which height I felt it was safe enough for me to disengage the autopilot. I completed my mission, during which I spent some time thinking about the young airman who had done the daily inspection of the instruments and had left the autopilot engaged. However, I should have done the proper cockpit checks which I certainly did ever after. My crew were never aware of my lapse and I am just as ashamed to tell this story today as I would have been had I told it 45 years ago. There are lessons of course: don't be distracted and don't assume — check.'

Yet to err in combat or under severe pressure is only human, and it would be wrong to imply that there was any widespread lack of professionalism in Bomber Command during the war. In fact, there were cases where men coped superbly even in the wake of 'finger trouble'. At 18.13hrs on 14 April 1945, Lancaster 460/S took-off from Binbrook to attack Potsdam. The outward trip was uneventful but 10min before target the Lancaster was intercepted by an unidentified aircraft from the starboard beam without any warning. The pilot's harness was shot away and the flight engineer, who was standing by the pilot searching the skies, was severely wounded. He fell across the pilot's seat and despite the navigator's ministrations, he died within a few minutes.

Immediately after the interception, the pilot took evasive action while calling for each member of his crew to check in. There was no reply from the mid-upper gunner, so the wireless operator was sent to check. He found that the mid-upper gunner's intercom, oxygen and turret hydraulics had been shot away, but the gunner himself was uninjured though covered in oil.

The captain now had to earn his money. His aircraft had been shot-up, his crew were scattered about the aircraft, and he was only 30 miles from target release. As the Lancaster was still to the west of the bomb line over Allied-held territory, he could not jettison his bombs without endangering friendly troops so he decided to press on with the attack. A quick scan of the dials showed the starboard inner engine to be unserviceable, but in trying to lean across to the engineer's panel and feather it in the dark, the pilot pushed the port inner feathering button by mistake. On seeing the engine wind down the pilot tried to unfeather but he now pushed the port outer button leaving the Lancaster flying on only one engine. In this condition, the bomb-aimer used his damaged bomb sight to drop within 4,000yd of the aiming point from 16,000ft. Immediately afterwards, the pilot unfeathered both port engines and continued on the ordered route.

On the way home, the navigator took over the engineer's duties as well as his own. His labours were hindered by the presence of the dead engineer and he tried twice to move the corpse in between everything else. But each time he became exhausted and it took an hour-and-a-half before the navigator and bomb-aimer — who had been pushing out 'chaff' over defended areas — were finally able to move the engineer to the rest bed. By this time the aircraft was down to 7,000ft so there was no longer any need to struggle about breathing oxygen. The bomb-aimer then took over the engineer's duties until the multiple-holed Lancaster staggered back to land at base four-and-a-half hours after leaving Potsdam. 'Throughout the whole occurrence,' concluded the post-flight report, 'the captain and crew revealed a high standard of discipline and consideration for our troops below.'

At the end of the war, Plt Off Prune was demobbed to return to his family home at Ineyne Manor, but unfortunately he left his spirit behind. The last flight of VX770, the first prototype Vulcan bomber, was a case in point. Built by Avro, the four-engined, delta-winged VX770 first flew in August 1952 powered initially by 6,500lb Avon RA3 engines and then by Sapphire SA6 engines of 8,150lb thrust. VX770 was used for aircraft handling and performance assessment until August 1956, but once production Vulcans entered service with the RAF, the prototype was deemed to have done its job. Over the next year, therefore, VX770 was converted to act as a flying test bed for the latest Rolls-Royce Conway bypass engines then undergoing development for use in the Handley Page Victor 2 and Vickers VC10. By September 1958, VX770 had amassed a total of 1,130 flying hours of which 627 had been with Conway 11s rated at 17,500lb thrust.

On 20 September 1958, a Rolls-Royce test pilot was authorised to fly VX770 on an engine performance sortie finishing with a flypast at the RAF Syerston Battle of Britain Day. Also on board were a Fairey test pilot acting as co-pilot, a Rolls-Royce engineering observer and an RAF navigator. They were briefed by Rolls-Royce's chief test pilot that the flypast was to consist of two runs over the airfield at 200-300ft between 250-300kt at 70-80% engine rpm.

The engine test having been completed, the pilot of VX770 called Syerston air traffic control at 13.35hrs giving an estimated time of arrival of 13.55. At 13.50 another call announced that the Vulcan was approaching from the west for a fast run at 250ft: the aircraft would then be banked to show off its delta planform to the crowd before going round for a low speed run and then back to Hucknall. Seven minutes later the Vulcan appeared as advertised and began a run up the main 25/07 runway at an estimated height of 80ft and a speed of 350kt. Cine film taken by two members of the public showed that when VX770 was passing the Control Tower, it started a roll to starboard and a slight climb. Within three-quarters of a second, a kink appeared in the starboard mainplane leading edge approximately 9ft outboard from the starboard engine intakes. This was followed by a general stripping of the leading edge, the breaking off of the starboard wing tip, and a general collapse of the main spar and wing structure between the spars.

51

Right:
Vulcan prototype VX770 breaks up during a high-speed flypast over Syerston airfield, Notts, on 20 September 1958.

Centre right:
Down she goes . . .

Bottom right:
. . . leaving only so much scattered wreckage.

Now enveloped in a cloud of fuel vapour, the Vulcan then went into a slight dive and began rolling to port which, at 45° of bank, increased sharply at the same time as the tail fin came off. The remainder of the starboard wing was now on fire and the aircraft continued to roll to port with the nose lifting until it was vertical. The port wing leading edge then began to crumple and catch fire. VX770 was now standing on its tail, travelling in plan form relative to the line of flight with the topside leading. The Vulcan was then lost to view in an intense fire, reappearing with its nose now pointing almost vertically downwards after having cartwheeled. The aircraft continued down in this attitude until the topside of the nose struck the ground just where the taxiway joined the end of runway 07. The port wing destroyed a fire/rescue vehicle and runway controller's caravan, killing three of the occupants and injuring the fourth. All four Vulcan crew members also died. The whole catastrophe had taken a mere 6sec.

The wreckage trail extended over 1,400yds along the line of flight, showing that the Vulcan had come apart in the following order: starboard wing leading edge, starboard wing upper surface, starboard wing lower surface, port wing leading edge, starboard wing centre section, rear fuselage, fin, main wreckage and finally the engines. Detailed examination of the starboard wing wreckage revealed that the leading edge had collapsed and become detached by rolling upwards and backwards over the wing front spar like the lid of a sardine can. Initial upward movement of the leading edge had started with the bending of the internal rib structure; once the upper external skin ballooned upward, allowing the lower ribs to bend further, the whole lot collapsed under increased bending loads. The port wing, being in turn subjected to heavy downloads in the air, then followed suit.

The powerful Conway engines were not to blame for any of this — over previous months the structure of VX770 had proved to be more than capable of coping provided that the aircraft was handled properly. However, VX770 was limited to a maximum speed of 380kt with a normal acceleration of 2.25g in straight and level flight; these limits had to be *halved* if aileron was applied for a rolling pull-out.

Analysis of the cine film showed that the Vulcan was clocking up 410-420kt as it sped down the Syerston runway. The aircraft also came within 65-70ft of the ground, which meant that the pre-flight brief was exceeded on two counts, but even so the Vulcan should have held together. Where the pilot completely overstepped the mark was in rolling to starboard while he flew so fast. He was probably trying to display the delta planform early, but in rolling to a rate of 15-20°/sec while also pulling 109,082lb of Vulcan into a rate of climb of

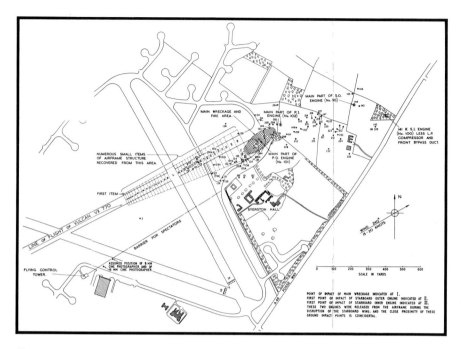

Above:

Map of Syerston airfield pinpointing the disintegration of Vulcan VX770. The spread of debris shows how widely wreckage can be thrown when an aircraft comes apart at high speed. Fortunately, no spectators were hurt at Syerston because they were all behind the crowd line which ran parallel to the display axis – imagine the mayhem if VX770 had run in *towards* the crowd.

3,000ft/min, he was imposing a strain of between 2 and 3g when he should have stayed below 1.25g. The airframe did not stand a chance and it broke up at the very point on the wing leading edge which Avro tests in 1950 had shown to be the weakest part of the structure.

In trying to show off the potency of his aircraft to the best effect, an experienced test pilot let his enthusiasm cloud his common sense and judgement to a fatal degree. But the picture is bigger than that. At least two of the other three unfortunates on board the Vulcan were all highly qualified aircrew, so they were in a position to counsel caution. Were they content to sit back and let the captain do as he pleased, secure in the belief that he had flown the Vulcan so often that he must have known what he was doing? Experience does not of itself grant immortality, but there is always a probability that it will be misinterpreted as infallibility. Whatever your aircrew specialisation on the flightdeck, you are never a passenger.

The saddest examples of 'finger trouble' or lack of awareness are always those which involve loss of life or aircraft for stupid or clearly avoidable reasons. Operating aircraft always demands particular care, especially when that aircraft carries weapons. The fog of war is still very dense at times, as the Arabs found in the 1973 war when they not only shot down 81 Israeli aircraft but also

69 of their own. But care must also be taken when simulating the pressure of combat in peace.

On 25 May 1982, a pair of Jaguars was recovering to base after a low level training sortie. As they travelled back in arrow formation at 1,000ft, the pilot of the No 2 aircraft was alerted by a ground radar unit to the presence of conflicting traffic. Suspecting that he was about to be 'bounced' for a training interception, the No 2 pilot scanned the skies and detected an aircraft passing down his starboard side. He then turned his attentions towards looking for a reported F-104 Starfighter in the front hemisphere and he had just identified it passing head-on beneath the pair when a loud explosion occurred. The No 2 Jaguar began to gyrate violently and on hearing his leader's call to eject, the pilot immediately pulled his seat pan handle. Fortunately he only sustained minor injuries and the main aircraft wreckage landed on open farmland.

Ten minutes previously, a Phantom was scrambled to mount a low level Combat Air Patrol in support of a fighter control station which was then being subject to a Tactical Evaluation. Because such Evaluations assess readiness for war, the Phantom was loaded with live missiles. Soon after getting airborne, the Phantom pilot completed his pre-attack checks but they were not monitored by his navigator who was preoccupied with checking his radar which had a history of minor unserviceabilities. The navigator then picked up a radar contact which he assumed would pass down their starboard side, and he informed the fighter controller that they would intercept it. As the Phantom turned in behind the contact, the pilot identified a pair of Jaguars flying in loose echelon formation which he decided would make two good exercise 'kills'. The pilot selected one of his infra-red Sidewinder missiles and the appropriate illuminations showed that the missile was ready for firing. Unfortunately, these indications were the same as those obtained from training missiles with which the Phantom force usually flew.

The Phantom pilot then aimed at the right hand Jaguar and obtained a lock-on signal from the Sidewinder. After ensuring that the Master Armament Switch was on and starting the gunsight camera to record the training 'kill', he squeezed the trigger in response to the navigator's 'In range clear to fire' call. The Sidewinder left the aircraft and the pilot could only watch helplessly as the missile homed inexorably on to the Jaguar's exhaust: the Jaguar was blown in two. An amazed F-104 pilot, standing off waiting for his turn to execute a practice intercept, thought to himself, 'My God, these chaps play for real'.

The subsequent inquiry found a series of pressures and oversights leading up to the accident, and like the links in a chain, any one of them could have been broken to stop the tragedy happening. For a start, the orders stated that live missiles should only have been fitted with safety locks, but the orders related to an older type of fighter missile and no Sidewinder safety locks had been cleared for flight. The Master Armament Switch should have been taped in the 'Safe' position, but when the navigator sent an armourer to get some tape he was told that none was available. The hectic pressures of the Tactical Evaluation also resulted in an element of confusion on the ground. Consequently, although the orders stated that the fighter controller should remind the pilot to check that his weapons were safe prior to closing in on the target, the controller did not do this because he had not been informed that the Phantom was carrying live weapons.

Above:
RAF Germany air defence Phantom sporting drop tanks, Sidewinder heat-seeking acquisition rounds and ventral gun pod. It was a lethal combination, just as capable of shooting down friend as well as foe.

Whatever the rights and wrongs of the background organisation that enabled this accident to happen, the facts are that the pilot was flying in a Phantom that had a camera mounted in front, that had no makeshift guard over the Master Armament Switch and which received no live weapon reminder from the fighter controller. These features were identical to the pilot's dummy weapon environment and, though he knew that his aircraft was fitted with missiles, he went ahead and shot down the Jaguar when intending to photograph it. Perhaps a piece of tape would have interrupted the pilot's automatic sequence and made him realise that an unintended course of action was being embarked upon. Yet at the end of the day, orders, instructions, locks and tape can only achieve so much: the final and most important safety break has to be the common sense and awareness of the crew themselves. At some stage in the simulated war, the Phantom crew lost touch with reality and became oblivious to the fact that their aircraft had live weapons rather than the dummies they carried on most occasions. Training for war must be as realistic as possible to achieve full operational capability, but playing hard rules demands special standards of care. Weapons such as Sidewinder are designed to be deadly: they must always be treated with respect. The loss of life or a multi-million pound aircraft are too high a price to pay for 'finger trouble' or lack of safety nous on the part of aircrew. The words of Capt A. G. Lamplugh, written over 60 years ago, are just as relevant today as they were then: 'Aviation is itself not inherently dangerous, but to an even greater extent than the sea, it is terribly unforgiving of any carelessness, incapacity or neglect'.

Below:
Jaguar GR1 (foreground), similar to that shot down inadvertently by an RAF Phantom on 25 May 1982. In the background, pulling away from a 'bounce', is a *Luftwaffe* Starfighter. The F-104G (for Germany) was an almost completely redesigned Starfighter insofar as Lockheed changed a light, simple fine weather interceptor into a complicated and much heavier multi-role combat aircraft primarily for the benefit of the *Luftwaffe* and *Kriegsmarine*. During the procurement process, much mud was slung at the F-104G but not even the most pessimistic competitor was prepared for the loss rate suffered by the Germans in their first few years. Eight aircraft per 10,000 flying hours were lost in 1961, rising to 13.9 the following year. In 1963 the rate fell to three, but this statistical improvement only came about because many more hours were flown – the actual number of lost aircraft rose. In 1964 the rate was 6.2 and a year later 8.45, or a Starfighter written off every 10 days.

Too many German pilots died in what the media soon dubbed 'The Flying Coffin', but it would be unfair to put too much blame on the aircraft itself for this sorry state of affairs. (The Royal Norwegian Air Force, which acquired the same variant, went almost 10 years before suffering a F-104G fatality.) Martin-Baker ejection seats were fitted after the 65th crash but the primary reasons for the Germans' troubles were human rather than mechanical in origin. For a start, German military air expertise stopped in 1945 and did not resume until 11 years later. This left a mass of young men with very little experience at one end, and at the other a lot of much older pilots who had plenty of World War 2 flying under their belts but little jet experience and certainly not nearly enough on second generation 'hot' ships such as the F-104.

The same lack of 'middle management' experience applied to engineers and other support staff, to which was added political pressure. The Luftwaffe was not only being given too many new aircraft at too great a rate for its own good, but it also felt under pressure to become as good as other NATO air forces as soon as possible. All efforts were put into meeting NATO declaration rates and if this meant that there were too few aircraft left to maintain proper flying training standards, there were insufficient good flight commanders around to cry 'Halt'. It was not the best way to help young pilots make a safe transition from the sunny skies of Arizona where they trained to the overcast and overbusy environment of NW Europe. Some Germans were also a bit too stiff-necked to heed good advice from other F-104 operators.

It took the appointment of the old wartime jet fighter man, Lt-Gen Johannes Steinhoff, to get F-104G servicing put on right lines and a proper pilot training programme initiated such that the loss rate dropped to an 'acceptable' level of 1-2 per 10,000hr. After 1970 the *Luftwaffe* was up and running properly, the bugs had been ironed out of the aircraft systems and experience levels were coming right. The F-104G was then revealed to be just another aeroplane, which only went to show that in military flying most particularly, it does not pay to run before you can walk.

Two Steps Forward, One Step Back

It is a truism that a *superior* pilot is one who stays out of trouble by using his *superior* judgement to avoid situations which might require the use of his *superior* skill. Yet it must be said that there are some aerial occurrences which even the most superior pilot can do little about other than to try and counter their effects as best he can.

On the afternoon of 20 October 1922, Lt Harold Harris walked out to a small high-wing Loening monoplane fighter standing in front of the hangar at McCook Field, Dayton, Ohio. Although only 27 years old, Harris was Chief of the Flight Test Section of the Engineering Division of the US Army Air Corps, and he was about to test the Loening to which experimental balanced ailerons had been fitted the previous day.

Standing beside the aircraft, Harris tried to clip on his parachute. A new seat cushion had been fitted to it that morning and as he struggled to fasten the harness, Harris found to his annoyance that the cushion had made the straps too tight. He called for another 'chute but its harness was even smaller than his own. Finally he decided to ignore Air Corps orders and fly without one. Although manual parachutes had been in service for more than a year, no one in the Air Corps had ever needed to jump out and anyway, if anything went wrong, he could surely get his machine down on the ground. But as Harris was about to climb aboard his aeroplane, he changed his mind. It was better not to tempt providence, especially on a test flight, so he strapped on his own 'chute and resolved to put up with the discomfort.

Five minutes after start up, the Loening was being levelled off in the clear sky at 2,500ft over Dayton. The pilot was tasked with making a 'manoeuvrability test' and on catching sight of another aircraft from McCook, the pair 'manoeuvred' to their hearts' content in mock combat. Suddenly as Harris got into position for a 'kill', his aircraft shook with tremendous vibration. The stick was snatched out of his hand and started slamming fiercely from side to side. Every time Harris made a grab for it, it was torn from his grasp.

He knew immediately what was wrong – the new ailerons had been set with too big an area forward of their hinges, and at high speed the forward pressure was overbalancing them. To regain control, Harris throttled back to lose speed but still the stick thrashed about. As he could now see that the surface of the wing was starting to ripple and that pieces of fabric were breaking away, he undid his safety belt, stood up in the cockpit, and felt himself plucked out by the slipstream. Once clear of the aircraft, he searched for the ripcord handle. Three times he pulled and three times nothing happened; he had been pulling the leg strap fitting. Finally he found the handle and at 500ft from the ground, the 'chute opened. Harris landed in a small grape arbour in the backyard of a house in Troy Street, Dayton, and as he disentangled himself from the vines, he had the satisfaction of becoming the first person ever to be saved in an emergency jump from an aeroplane with a manually operated parachute.

If jet engines brought about new features such as swept wings and nose wheels, they also enabled pilots to fly much higher and faster than had ever been possible under piston power. This in turn heralded a new set of threats to life and limb, and to prove that experience does not of itself confer immunity from harm, 32 test pilots died in the UK between 1945 and August 1951. Many of them had no chance of surviving accidents which resulted from aircraft design problems or technical malfunctions because the manually operated parachute had distinct limitations when it came to escaping from aircraft capable of speeds in excess of 400mph.

On 4 January 1944 a Royal Aircraft Establishment (RAE) test pilot, Sqn Ldr Douglas Davie, took-off from Farnborough in DG204, a prototype of the Meteor twin-engined jet fighter which operated under the codename 'Rampage II'. The weather that day was very good with very little cloud and slight heat haze. Heading northwest initially, Davie aimed to carry out an altitude performance climb to 20,000ft and then explore the stalling characteristics of DG204's Metrovick axial-flow engines at 30,000ft. The whole flight should have taken an hour and as Farnborough employees returned from lunch around 13.50hrs, they could see the silvery machine high above heading back towards the airfield. Suddenly DG204 fell apart and dropped to the ground in pieces.

Witnesses varied widely in their estimates of the height at which 'Rampage II' broke up, but once the main wreckage was collected it became clear that DG204 had reached around 15,000ft when the port engine suddenly and completely disintegrated. The aircraft then entered a dive whereupon Sqn Ldr Davie got out, but his arm was broken getting rid of the canopy and then the arm was torn off. He then struck the tailplane with some force so when he finally fell clear, it was highly unlikely that he was in any fit condition to pull

Below:
The remains of *Rampage II* having come to rest inverted after the prototype Meteor fighter broke up over Farnborough in January 1944 . . .

Above:

. . . the tail came to rest separately on the roof of a Farnborough shed.

the ripcord. Sqn Ldr Davie died as he fell through the roof of a small building just to the south of Farnborough's famous Black Sheds.

The starboard engine carried on running in the descent and between 7,000-10,000ft, the complete tail section fell off finishing up on the RAE foundry roof. The remaining mass then fell almost vertically to the ground, landing inverted; lighter pieces were distributed up to 13 miles south of Farnborough by very strong upper winds.

Accident investigators had now to determine why the port engine came to grief. Search parties were despatched along the 13-mile wreckage trail but it took a year of scouring fields, built-up areas and even a military sewerage works to find all the engine components. Nevertheless, as early as March 1944 the Board of Inquiry had discovered enough to state that, 'the Primary cause of the accident was the port engine failure, due to the bursting of the compressor rotor'. The bursting resulted from a combination of overspeeding and structural weakness, and in an effort to prevent a repetition of these technical defects, the Board recommended that the rotor structure be strengthened, inspection procedures be tightened up, and the possible causes of engine overspeeding be examined to see if there was a need for a more positive governor between the fuel pump and compressor rotor.

It is always the same when aircraft push back the frontiers: there are new technologies to absorb, new tolerances to be applied and new techniques to learn. Thus, when the DG204 Board found that the 'rear fuselage failure was probably due to the speed and yaw the aircraft attained in the subsequent dive', they were not criticising the original aircraft design. All available information suggested that the strength of the tail unit was adequate to meet all normal flying loads, but when the accident occurred, unusual and rare loads were encountered which removed the tail. Gloster designers could only strengthen the tail structure in response to a situation which none of them could have been expected to predict before the tragedy happened.

The transition from piston to jet age forced pilots to adapt to a whole new way of life. Take the engines themselves. On 27 January 1943, Gloster's chief test pilot Michael Daunt was standing by the side of a prototype Meteor's engine nacelle. The engine was turning because Daunt wanted to check if the joints of the fuel pipes were leaking. He then stepped back from the side of the engine practically in line with the intake, whereupon he was sucked bodily off his feet inside. 'My ribs did a bit of tin-bashing and it was very noisy. My first reaction of being inside that air intake was to hold my breath, as I was quite convinced that if I didn't I could have a collapsed lung. I have never ceased to be amazed how quickly the human brain reacts in such situations.' Luckily, the man in the cockpit closed the throttle almost immediately and a shaken Gaunt was pulled out. He was the third British thing ever to be sucked into a jet

Below:
A live test of an early Martin-Baker ejection seat from a specially modified Meteor T7. *Martin-Baker*

engine, the previous two being a hat and tea-tray, but he was not to be the last. A moment's carelessness in front of today's powerful gas turbines is likely to have much more tragic consequences.

In an effort to get jet aircrew safely away from a stricken aircraft, the Germans made the first successful use of an ejection seat on 13 January 1943 when a pilot in a He280 jet fighter experienced severe icing and lost control. Resourceful wartime German engineers produced several types of seat employing compressed air or an explosive charge or a large spring or hydraulic power – in fact just about every power source available apart from a huge rubber band. But Germany dropped out of the advanced aeronautics race in 1945 and it fell to Martin-Baker to take over pole position when Bernard Lynch ejected from a specially modified Meteor on 24 July 1946.

The idea of being catapulted into space on what was roughly equivalent to a 37mm cannon shell did not immediately endear the 'bang seat' to some aircrew, but time was to prove its worth. The most convincing test came during the period when there was considerable interest in the potential of small rocket motors for boosting military aircraft to altitudes well above those attainable with air-breathing engines. One such development was Napier's Scorpion liquid rocket motor, and double Scorpion installations were made on two Canberra B6 bombers taken from No 76 Squadron at Wittering. These aircraft were then used successfully to sample the mushroom cloud over the Christmas Island thermonuclear tests, but in early April 1958, WT207 suffered a rocket motor problem at extreme altitude over the UK. The crew of Flt Lt de Salis and Flt Lt Lowe ejected successfully around 56,000ft. Ejection seat automatics are designed to operate the main 'chute only on reaching denser lower air, so both men free-fell to 10,000ft, suffering acutely from the cold which was initially 97°F below zero. De Salis' stabiliser 'chute hit something as he ejected and for 'several minutes' he was semi-conscious in a wild spin with arms flung out above his head. Lowe fell face-down for what 'seemed a day-and-a-half', but eventually both their main canopies deployed on cue. The important point was that both men survived with only minor injuries and some frostbite from the greatest height to date from which a crippled aircraft had been abandoned.

It was just as well that Martin-Baker ejection seats were to prove so effective because there was to be a great need for them. The Meteor F8 for example, which was to be the RAF's primary fighter between 1950-55, had an unenviable accident record by modern standards. To take just one instance, on 18 June 1951, No 615 (County of Surrey) Squadron of the RAuxAF was being inspected by its honorary Air Cdre, Sir Winston Churchill. During the flypast, one aircraft got into difficulties and crashed into the garden of a nearby bungalow. Two other pilots, distracted by the resulting fireball, then collided bringing the total to three pilots killed and three aircraft lost.

On 21 December 1954, No 257 Squadron launched a four-aircraft formation of F8s from Wattisham. The sortie included a tail-chase with the Meteors spaced out some 400-600yd astern of each other so that their pilots could carry out ranging and tracking practice on the aircraft ahead. After a series of high speed manoeuvres, the No 2 pilot's speed increased to Mach 0.82 whereupon his aircraft went out of control. This was not unexpected because the F8 was known to encounter severe compressibility followed by loss of control above

Mach 0.8 at height. Standard recovery procedure was to select airbrakes out and throttle back, for once the speed decayed, control could be recovered. So the No 2 pilot did all the good things, only to find that his Machmeter stayed fixed at Mach 0.82. Descending vertically through 10,000ft, he decided that it was time to leave. He jettisoned the hood but the high 'g' plus severe turbulence prevented him reaching the firing handle above his head. The pilot's attempt to push his right hand up into the slipstream with his left hand only resulted in a broken wrist, but a further try using just the left hand proved successful. His legs were then splayed by the force of the slipsteam such that a leg and ankle were broken and four toes were ripped off his right foot, but the pilot was lucky to survive at all because witnesses said that his seat fired at around 1,000ft which, given the rate of descent, must have been outside the seat's design envelope.

During the subsequent inquiry, it became clear that there was a marked similarity between this accident and other previously unexplained fatal crashes when Meteors were known to have spiralled into the ground from height. It was found that recovery from the effects of compressibility was much more difficult if aileron plus 'g' loading were being applied at the time. Henceforward, Meteors were limited to Mach 0.8 at all levels and warnings were given to pilots about the dangers of pulling 'g' and applying aileron at high Mach numbers above 20,000ft. The military also learned that it was wise to fit a firing handle between the knees and leg restrainers to all ejection seats, but in the absence of an automatic canopy jettison system, it became common practice for Meteor pilots to eject through the hood. As the year progressed, Martin-Baker ejection seats became more and more capable, and eventually rocket-propelled seats would save lives even when initiated at zero feet and zero speed. But no matter how good an ejection seat is in theory, it is only so much expensive metal if it is not used in time.

As single-seat Meteor fighters entered service around the world, it became clear that a two-seat training version was highly desirable to give pilots their first experience of jets after initial training on piston types. To save time, Glosters basically added a new and longer nose to their F4 fighter and at a stroke created the Meteor T7 trainer. The RAF took delivery of its first T7 in December 1948 and thereafter T7s went to the Advanced Flying Schools (AFS) beginning with No 203 AFS at Driffield in East Yorkshire.

Almost at once the AFS started to experience a high incidence of the same type of accident. Meteor T7s returning to the circuit would be making their way towards the end of the downwind leg when suddenly they speared into the ground. Stories of the ever-enlarging hole which marked the spot at Driffield lost nothing in the telling, but because ejection seats could not be fitted into the hastily expanded nose of the T7, there were no survivors initially to brief the accident investigators and therefore the phenomenon became known as the 'Phantom Dive'. It was also no respector of rank or experience: instructors and students alike suffered its effects and even an Air Vice-Marshal commanding a fighter group died whilst visiting one of his stations.

The basic cause of the 'Phantom Dive' was eventually found to be loss of direction control. It was confined to the T7 because the addition of the large canopy brought the centre of lateral area further forward than on single-seaters.

Above:
The production Meteor T7 trainer's large canopy accommodated pupil and instructor in tandem but in so doing brought the centre of lateral area further forward than on single-seat variants.

This effect was exacerbated when a ventral tank was added, and directional stability was only maintained because the fin gave more lift at small angles of attack than the forward fuselage side surfaces. However, anything which reduced the effectiveness of the fin reduced directional stability and if yaw was then induced, it could easily lead to loss of directional control. The two easiest ways of reducing fin effectiveness were to fly at high angles of attack, such as low speed in the circuit, which blanketed the fin, and selecting airbrakes out to induce turbulent air over the fin. Combining the two spread the turbulent airflow higher up the fin and if the undercarriage was down, the nosewheel added to the problem by presenting a side area well ahead of the centre of gravity.

Thus as he started his downwind leg, the T7 pilot might have his airbrakes out to help the speed decay to 150kt. Directional stability was now much reduced but it was manageable unless the pilot forgot to retract the airbrakes before selecting undercarriage down. The port wheel always lowered before the starboard and if rudder was not applied, the fin could not produce any worthwhile lift on its own to oppose the yaw, especially if it was being affected by the airbrakes. So the forward side surfaces took over to increase the divergence, the fin stalled, the increasing yaw induced a powerful roll to the left, the nose dropped and the aircraft had entered a 'Phantom Dive'. The best recovery method was to roll into the spiral, retract airbrakes and undercarriage, and pull out, but even if you had your wits about you this could take nearly 3,000ft. If you were at 1,000ft in the circuit when the 'Phantom Dive' crept up, there was nothing to stop your fall but the ground.

'Phantom Dive' was the classic case of an unforeseen problem being built into an aircraft redesign, and it was only eradicated as ab initio pilots had it drummed into them to select airbrakes in before lowering the undercarriage, to correct the yaw while the undercarriage travelled, and to keep the sideslip ball central particularly in the circuit. Perhaps such killers have gone for good now that aircraft designers have more time and computer assistance at their disposal, but there is a limit to the forgiving nature of any aeroplane if it is not handled properly.

There is a postscript to this particular story. On 30 May 1988, the RAF's last flying Meteor T7, WF791, took-off from its home base so that the pilot could display the old stalwart at an air pageant being held at Coventry airport. The weather was good and the display sequence was very restrained. The pilot's normal sequence had lasted for about 3min when he did a wingover to the right, intending to bring the Meteor back along the display line with both undercarriage and flap extended. The undercarriage appeared to lower normally as the pilot climbed the T7 to the highest point of the wingover when the speed was assessed around 150kt. As the aircraft began the descending turn back towards the airfield, the roll rate appeared to be faster than on previous occasions, the bank increased to 45° and the nose dropped. The Meteor then turned rapidly through 90°, settled into a dive and crashed in open ground close to the airfield. The Meteor was destroyed and the pilot died on impact.

It became clear from a spectator's video recording that WF791 had been flown throughout the sequence with the airbrakes extended. This was contrary to normal practice, verbal warnings, and written advice in the *Meteor T7 Pilots' Notes*, no doubt caused by a mental aberration on the part of the pilot. The Board of Inquiry had to conclude that the 'Phantom Dive' had struck again. The very experienced and highly qualified pilot concerned was used to flying much more modern machines, and perhaps modern technology has now

Below:
Meteor T7 photographed immediately prior to impact at the Coventry air show in May 1988.

become so good that it can delude us into thinking that the old perils of flight are no more prevalent today than highwaymen on the roads. Following the sad loss of both WF791 and its pilot, it is worth remembering that even old aircraft can still bite.

In the 1950s, many accidents could be explained away by ignorance. Asymmetric flying with one engine shut down had been no great problem on piston-engined twins, but the higher thrust available on the Meteor generated a high workload in maintaining control at low speed such as during an engine failure after take-off. Asymmetric overshooting had to be undertaken with equal care and 'rollers' on one engine were only to be recommended in dire emergency because if the throttle was opened too quickly, the resulting yaw could be uncontrollable. Eventually the RAF evolved an excellent training syllabus to prepare pilots for the pitfalls of asymmetric flight, but it would take quite a few deaths to prove that the new jet age was no place for the uninformed or the unprepared.

In the first month of 1954, the RAF lost 41 men in 14 accidents: December was not much better with 20 fatalities from 10 accidents, and these figures do not include crashed aircraft whose crews lived to tell the tale. The accident rate did not improve appreciably for some years, not least because some accidents resulted from circumstances beyond any pilot's control.

The Supermarine Swift for example was rushed into service in 1954 because the RAF feared it was being outperformed by almost every major air force in the world. However, Britain's latest high altitude interceptor was restricted to a maximum altitude of 25,000ft because it had serious control problems above this height. The Swift F1's engine could surge with or without gun firing, and its electrically actuated tailplane had an alarming tendency to 'runaway' after take-off. Following two fatal accidents, the Swifts were grounded pending system checks; several weeks later they were cleared for limited flying, which was again suspended after two more fatal crashes. Supermarines tried to improve its creation but the Swift's accident rate grew as high as 21 per 10,000 flying hours. The Swift was soon phased out of service as a fighter, reappearing at a later date in the fighter-recce role.

The Swift's contemporary, the Hawker Hunter, also had its problems in the beginning. From 1954, experienced test pilots expressed doubts about the Hunter F1's performance, varying from complaints about severe canopy misting during the descent to four instances of engines flaming out during gun-firing. The first RAF squadron to receive the type — No 43 Squadron — suffered two accidents following flame-out, though fortunately neither proved fatal.

The superior pilot could only do so much if his solitary engine failed. On 4 February 1957, Flt Lt A. W. Picking of No 71 Squadron took-off from Brüggen to fly Hunter XF362 on an acceptance air test.

Right:
Supermarine Swift FR5 after coming to grief at Benson, Oxon, at the end of October 1958.

'At 45,000ft I throttled back to 7,800rpm and carried out tight manoeuvres between Mach 0.75 and 0.95. The aircraft handled satisfactorily apart from wing heaviness with trim zeroed. By this time I was approximately 40 miles south of base at 28,000ft, so I throttled to approximately 7,200rpm and headed for base in order to check the stalling and manual flying characteristics. I flew straight and level at Mach 0.85 for approximately three minutes. Suddenly there was a violent explosion and severe vibration which stopped when the engine had completely run down. The fire warning light did not operate. I immediately throttled back but did not switch off any controls as I had both hands near the ejector blind handle. When all was quiet, I tried the flying controls which responded satisfactorily and then selected 'manual' as the hydraulic pressure was almost zero. I next switched off the High Pressure and Low Pressure cocks, booster pumps and all other electrics except the battery master and radio. At this stage (09.42hrs) I made a 'Pan' call on the approach frequency and arrived overhead base at 18,000ft. The only handling trouble I experienced was bad shaking and vibrating from the rear and whenever back pressure was applied to the control column. The standby tail trim would not work.'

Picking's CO, Sqn Ldr E. M. Sparrow, was airborne at the time and on hearing the 'Pan' call he immediately headed for Brüggen overhead where he joined up with XF362 and another No 135 Wing Hunter being flown by Lt Maina. Maina managed to get in close and was able to pass Picking details of the damage to the rear and centre fuselage. Picking was keen to know if the rear end was burning:

'. . . on hearing that this was not the case, I decided to stay with the aircraft. At 12,000ft I checked the stalling speed which was 130kt and flew one complete orbit to position myself for a landing on runway 27: had I misjudged this, I could have repositioned for a landing on runway 09. I do not remember my

height after the orbit but the position looked favourable so I flew across the downwind leg and selected wheels down opposite the caravan. This gave one red light so I used the emergency system which at first would not operate because I had forgotten to press the knob before pulling the handle. I commenced my final turn with "three green lights" at approximately 4-5,000ft and selected 10° of flap normally. Owing to excessive height on finals, I made several "S" turns and did not select emergency flap because at this time I required both hands on the control column.'

Sqn Ldr Sparrow watched it all with satisfaction:

'In his landing run Picking allowed the aircraft to roll to the end of the runway before steering it carefully clear on to the Operational Readiness Platform, thus completely clearing the landing area and not obstructing other traffic in any way. Throughout the whole of the emergency, from the time he experienced the engine failure, committed himself to a forced landing despite the knowledge of the damage to his aircraft, to the time he stopped the aircraft well clear of all aerodrome traffic, Flt Lt Picking demonstrated flying ability of a very high order, extremely clear and logical thought, and exceptional airmanship.'

One of the problems with early jet fighters such as the Meteor and Hunter was their engines' healthy appetite for fuel. Neither the Hunter F1 nor F2 had wing tanks or provision for drop tanks, and at least two pilots were killed due to fuel starvation during the F1's first year in service. Yet despite the known limitations, not a few pilots still insisted on ignoring them.

Towards the end of a flying training sortie in early March 1952, an instructor and his pupil in a Meteor T7 found themselves over Flamborough Head with only 100gal of fuel remaining. The instructor shut down the port engine to conserve fuel but even so there was very little sloshing about in the bottom of

Below:
Hunter XF362 showing the evidence of catastrophic engine disintegration after Flt Lt Picking demonstrated flying ability of a very high order to get it down on the ground safely.

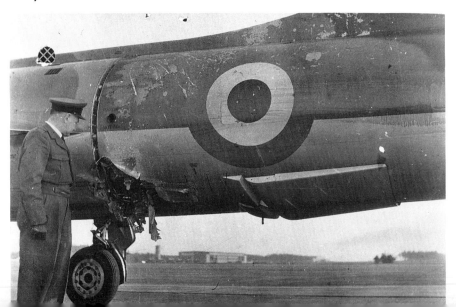

the tanks as the T7 approached Full Sutton airfield. The short 2,700ft runway was well placed to starboard and the instructor made a beautifully judged asymmetric approach given that there was no headwind component. Unfortunately, there was a railway embankment just short of the touchdown point, and what should be passing along at that moment but a fish train from Hull. The Meteor's nose went between two wagons, throwing several trucks off the rails, and the aircraft came to rest with the cockpit intact but swimming in over-ripe fish. A chastened crew was extracted with no more than a broken arm, dented pride and stinking flying overalls, but they had learned the hard way that it pays to know your aircraft limits and to stick to them.

Yet the most legendary case of trying to fly on the fumes occurred within the Hunter force in 1956. The saga began at West Raynham, Norfolk, home of the RAF's Central Fighter Estabishment (CFE). Half of CFE comprised the Day Fighter Leader School (DFLS) which was established after the war to teach present and future leaders of fighter wings and squadrons how best to fly their formations to their limits. To quote from the syllabus, a good fighter leader 'must know how to best operate and manoeuvre formations of aircraft in the face of varying tactical contingencies in order to maintain maximum cohesion whilst making the minimum concession in speed, climb and manoeuvrability'. In other words, DFLS saw itself as the elite who taught the embryo elite, but there is always a risk that any self-proclaimed master race will make minimum concessions once too often.

On 8 February 1956, two instructors and six students took-off at 10.50hrs in eight Hunters to carry out a four-v-four dogfight at 45,000ft. There was some low cloud and widespread mist but although the weather was expected to improve, at the time it was so foul that the other half of CFE — the Air Fighting Development School — refused to fly at all. The DFLS Hunters climbed to height, formed up into battle formation, made a few turns and then had to return to base because their fuel was running low. By the time they were in the West Raynham overhead at 20,000ft, they found that far from the weather improving, the cloud base was down to 400ft and fog had reduced surface visibility to less than 1,000yd. With some 20-25min of fuel remaining, the octet could have diverted either to Marham — less than a dozen miles to the southwest — or to Waterbeach, 20 miles further on in the same direction. A Meteor and Vampire had already successfully diverted from West Raynham to Waterbeach but it was decided that the Hunters would opt for Marham. Once over West Raynham at 2,000ft, the Hunters set off for their diversion with 30sec intervals between pairs.

It has to be said that being on a Canberra station, the controllers at Marham might not have been experienced in dealing with a bevy of fighters short of fuel. Their ground-controlled approach radar was also not of the best, but Fate's trump card was to roll the fog over Marham while the Hunters transited the short distance from West Raynham. Red One and Red Two overflew Marham at 1,000ft, made a quick circuit, broke cloud at 500ft and then flew into the fog. The instructor in Red Two lost his leader but pressed on, landing at 11.32hrs after 42min in the air. Red One flew three timed circuits before he saw the runway; he then landed successfully though his engine stopped from fuel starvation as he taxied in.

Two aircraft had been saved but that was to be all. Yellow Three could only snatch glimpses of the ground from 500ft so he climbed and ejected. Yellow Four pressed on but the squadron leader pilot on board WW635 died when his aircraft crashed into a field. Yellow Two — the other instructor — lost his leader and with 12gal remaining, climbed and ejected. Yellow One was slightly luckier. He was forced to climb when he saw trees in his path but on letting down again to 150ft for a low level circuit, he saw the runway. Unfortunately, his engine flamed out as he turned on to finals and he force-landed straight ahead. Red Four likewise lost his leader and on descending to 600ft without seeing the ground, he climbed and ejected as the engine flamed out at 2,500ft. Red Three also climbed to 4,000ft and ejected, his aircraft hitting the ground at 11.40hrs. In the space of 50min, the RAF had lost six aircraft and one life for no good reason at all.

The short time frame of this multiple accident said much about the internal fuel capacity of the Hunter F1 but the blame for the catastrophe on 8 February 1956 did not lie with the aeroplane. DFLS lost six Hunters that day because it had got into the habit of regarding normal safety procedures as constraints which only applied to lesser mortals, and they did not learn otherwise until it was too late. Flying to the limits is very commendable but flying beyond them is not. Once drop tanks were fitted, the Hunter became a superlative aeroplane, but despite the passage of time new aircraft can still enter service with limitations and bugbears that need ironing out. Therefore, even if you think you are operating an aeronautical sow's ear, fly the beast as it is rather than as you hoped it would be until the designers have time to turn it into a silk purse.

It is a sad fact that over-confidence and even arrogance have killed more pilots than faulty equipment. In the opinion of that exceptional US aviator and the first human to go supersonic, Brig-Gen Chuck Yeager, they were the undoing of Dick Bong, 'our top war ace in the Pacific, who became a test pilot. Dick wasn't interested in homework. He crashed on take-off when his main fuel pump sheared. He had neglected to turn on his auxiliary pump because he hadn't read the pilot's handbook, so he flamed out only 50ft up. He had no ejection seat, but stood up in the cockpit, popped the canopy and then his 'chute. The air stream wrapped him around the ship's tail, and he went in with his aeroplane.' It never pays to treat any aircraft with disdain.

An ejection seat should therefore never be regarded as a cop out for ignoring the correct drills or for not knowing the limits of the aircraft. Despite the challenges of new technology, too many accidents over the years have resulted from mundane error rather than from people being at full stretch. The RAF's first Canberra accident occurred when WD895, an early production aircraft delivered to Hawarden Maintenance Unit on 8 April 1952, was flown the following day by the unit commanding officer with the ferry pilot checking him out from the jump seat. During the landing checks, the check-pilot called for 'All fuel cocks on' whereupon the pilot under instruction switched them all off. As silence descended, an attempt was made at an engineless belly-landing on a grass airfield nearby, but the Canberra slid off across the boundary and one of the RAF's latest shiny bombers was written off. As Chuck Yeager put it in his usually succinct fashion, 'the best way to fly safe is to know what the hell you are doing'.

A Display Too Far

Two thousand years after the gladiators, there is still something in human nature that finds almost immoral excitement from watching other people taking risks with their lives. Air forces have long understood that the more thrilling an air display, the bigger the crowds, and big crowds mean good public relations.

In Britain today there are some 700 air shows a year, the largest of which draw two or three times the capacity of Wembley Stadium. But when talking of 200,000 or 300,000 people, it becomes harder to give them all a good view while keeping them out of harm's way. It is here that premier aerobatics teams such as the *Red Arrows* display their great skill. They are not just showmen; they are illusionists who convey an impression of danger while in fact maintaining the strictest safety standards. For example, the heart-stopping low level pass by the Arrows' Synchro Pair is separated laterally by the width of the runway, and despite appearances to the contrary it is no more head-on than cars travelling in opposite directions on a motorway.

But it is one thing to display as a permanently constituted team flying tried and tested aircraft — it is another altogether to publically push an experimental prototype to its limits. That takes a special sort of pilot, someone with 'the right stuff'.

There were quite a few superlative pilots around immediately after World War 2, and when some of the best found that they could not adapt to the constraints of a peacetime air force, they exchanged the excitement of combat for the spice of test flying. It was certainly a very interesting time to be a test pilot because the advent of the jet engine combined with the fruits of wartime aeronautical research were pushing back frontiers everywhere. Nevertheless, in the opinion of not a few experts, one frontier that could not be overcome was the 'sound barrier'. When Britain's first swept-wing jet, the DH108, disintegrated around Mach 0.94 at the hands of Geoffrey de Havilland Jnr on 27 September 1947, it only confirmed the fear that the 'sound barrier' was an invisible wall of air that would smash any aircraft coming up against it.

The DH108 test flying programme was taken over by John Derry, an ex-CO of the Tempest squadron at the RAF's Day Fighter Leader School. He found himself flying the third prototype with its more powerful engine, but there was much more to successful high-speed flight than just sheer power. At a time when no wind tunnel measured beyond Mach 0.85, a series of patient and painstaking steps, each one bringing back a little more information on pitching and oscillation, were the keys to keeping both man and machine in one piece as they tentatively approached the limits of controlled flight.

John Derry piloted the DH108 to the 100km closed-circuit speed record in April 1948, thereby earning himself the Segrave Trophy for the greatest annual performance on land, sea or in the air. But this was just a little light relief, and all his meticulous research work was only fully justified in September 1948 when he became the first pilot in Britain to fly faster than sound. Derry was also an exceptional aerobatic pilot. His needle-sharp aerobatic displays found a

71

ready following among aviation enthusiasts, and the French were particularly won over by the good-looking as well as outstandingly able 'Le Blond' English pilot. The year 1949 saw John spending a lot of time over France proving the worth of the Vampire V, and it was the French who coined the phrase 'le Derry turn' for the impressive low-level turn followed by a roll underneath and an equally tight reverse turn which John introduced into the aerobatics manual and which few pilots could copy accurately.

When he eventually came to set down his thoughts on display flying in a 1952 *Aeronautics* article entitled 'Analysing the Art of the Demonstration Pilot', John Derry stated that a good pilot should aim both to entertain and to inform his audience. The proportions of entertainment and information varied according to the type of audience, which he divided as follows:

'There are the people who go to judge the show with a critical and experienced eye. There are those who, though not connected with aircraft, get genuine enjoyment and excitement from the sight of polished formation aerobatics or the erupting passage of a high-speed prototype.'

Finally, there were those who go 'because they consider these as circus acts with no safety net'. Aerobatics reflected a fighter's manoeuvrability, and Derry wrote that any sequence should be executed smoothly:

'Too often, each run over the crowd is punctuated by a pause and rush, rather like the short-winded bowler who takes a very long run. In a good display the aircraft is not performing a series of separate manoeuvres but one continual action which holds the attention from beginning to end.'

But while aircraft should be flown fast enough and low enough to astonish and amaze, Derry had no time for those who sought gratuitous danger. He certainly had no wish to emulate the French aerobatics ace Rozanoff who specialised in 'daisy cutting' high-speed runs and eventually died in the process.

'There may be some who prefer a dangerous looking show, but the best advice to those people is to stick to the fairground. Air displays, or at any rate demonstrations of different types of aircraft, are not intended to appear hazardous. An individual item which looks unsafe is usually badly executed and is often less safe than one at a lower height but skilfully directed.'

Trials on the DH108 came to an end in the summer of 1949 once the company decided that they had learned as much as they could from it and, after displaying various marks of Vampire, Derry progressed to the prototype Venom. This in turn led on to what was to be de Havilland's final military programme, the DH110. This large, swept-wing, twin-jet, high-speed interceptor was the heaviest, fastest, most advanced and costliest fighter de Havilland had conceived, and it was in fierce competition with Gloster's GA5, later to become the Javelin, for a big RAF contract.

The first DH110, WG236, was rolled out at Hatfield in the summer of 1951, and it first flew at the hands of company chief test pilot and ex-wartime nightfighter ace, John 'Cat's Eyes' Cunningham, on 26 September. Once the

Venom test schedule finished at the end of that year, Derry became responsible for the 110's flight-test programme, getting airborne in the big silver beast for the first time on 22 January 1952.

Until the end of February he flew alone on general handling, but from then on he flew on nearly every occasion with Tony Richards as his flight-test observer. WG236 became the first two-seat, twin-engined aircraft to break the 'sound barrier' with John and Tony on board on 9 April, and periodically thereafter the peace around Hatfield was disturbed by a shattering double sonic boom. But the air around WG236 was not the only thing to be disturbed by this time. During the summer of 1952, John Derry decided to leave de Havilland. The pay was not all that good and he knew that the company was unlikely to design another military aircraft. De Havilland's future seemed to be much more bound up with civil types such as the Comet, and as long as he was engaged in experimental flying, Derry wanted it to be with military aircraft. But the main cause of his frustration was that he could go no further at de Havilland. With John Cunningham firmly in the chief test pilot's chair, Derry saw no prospects of any further promotion. Moreover, he had the uncomfortable feeling that the company was expanding to such an extent that the personal touch was being squeezed out. Flying was no longer fun, and it seemed a good time to go.

He put out feelers for new employment and ironically came up trumps with de Havilland's great competitor in the interceptor race, Glosters. Negotiations went well and it was tentatively agreed that Derry would eventually take over as Gloster's chief test pilot after working for a while as Bill Waterton's number two. However, he decided to postpone any move until after the Society of British Aircraft Constructors' Display at Farnborough in September. 'I'll see de Havilland through Farnborough before I settle anything', he told his wife. 'It's only fair that I do — and I'll sell that plane for them, even if it kills me.'

The second DH110 prototype was ready in mid-July, during which time the Derry family were on holiday in Cornwall. It was a good opportunity to unwind, but the long journey back with a caravan in tow was exhausting. No sooner had John set foot back inside the door than Hatfield rang to say that he was to start test flying the second prototype immediately to amass enough flying hours for it to be allowed to give a full display at Farnborough. In not the most contented frame of mind, Derry set off for his first flight in WG240.

WG240 was soon to be painted satin black, a finish held to be more fitting than silver for an all-weather nightfighter. As the test programme progressed, Derry realised that the second prototype was much lighter than its predecessor with more powerful Rolls-Royce Avon RA7 engines to boot. 'Acceleration is out of this world', he enthused even though the starboard engine had an annoying tendency to run hot. Nevertheless, flight-testing the most advanced military machine in Britain was a long and arduous business, and with no advanced computers, fool-proof wind tunnels or sophisticated telemetry to take some of the load, Derry and Richards often ended up drained after long periods of concentration. On one occasion, both engines flamed out as an outside loop starved them of fuel. It was little things like this, combined with the strain of working up to a Farnborough performance designed to see off Glosters, that built up the pressure. Derry's wife noticed how tired and drawn he was looking,

Above:
The second prototype DH110 (WG240), the lighter and more powerful sister of WG236, on a test flight shortly before the 1952 Farnborough Show. *BAe*

and he had a new found tendency to snap irritably at innocent remarks. He was sleeping badly and living off his nerves, neither of which was conducive to peace of mind.

On Saturday 30 August, Derry and Richards took-off from Hatfield for the 20min flight to deliver WG240 to Farnborough. Today, the world's aviation companies are largely faceless corporations staffed by men who are unrecognisable to the general public. But in the immediate postwar years, larger-than-life aircraft manufacturers such as Sir Frederick Handley Page and Sir Geoffrey de Havilland were still at their desks, and their new and extraordinary shapes which appeared in the sky every year captivated the general public. The early 1950s were the heyday of Farnborough displays. It is hard to appreciate the excitement that early supersonic flights generated, but as Britons emerged from the slough of postwar rationing into the new Elizabethan age, they flocked to the annual Farnborough Show in droves. This was a time when man was striking out in all directions for the seemingly unattainable, whether by aiming for the stars, climbing Everest or running the sub-4 minute mile. Thus there was tremendous national pride in Britain's aeronautical achievements, and by 1952 the public was well aware that aerodynamicists were moving into regions where the behaviour of aircraft was still superficially understood, and that pilots were doing tests that were more hazardous than usual. It is no exaggeration to say that men like John Derry, or Hawker's chief test pilot Neville Duke, were national heroes because in making a reputation for their aircraft, they also made one for themselves.

One of the great features which interested and excited spectators in 1952 was to hear a double sonic bang as the aircraft approached the airfield to begin its display. It was possible to aim these bangs, and the great display pilots often took-off a few minutes before they were timed to perform to climb to height. They then dived — aircraft in those days were not capable of straight and level supersonic flight — until Mach 1 was achieved, and then they pulled out and descended into the circuit to give the remainder of their display at closer range.

John Derry understood that the experienced pilot, like any great entertainer, should make the most of his arrival:

'In a show lasting five or six minutes, the pilot cannot waste any time and if he wishes to include some aerobatics it may profit him to make his fast run as part of this sequence. On the other hand, if the pilot considers it of sufficient importance to get the absolute maximum speed of his aircraft, he may deem it worth giving away one or two minutes to achieving an accurate run-in . . .

'. . . the attitude of the aircraft, the direction of the run, use of a following wind (this is surprisingly effective), unexpected arrival, and the use of trees and buildings as a backcloth to give the maximum relative motion, all are points which are within the pilot's control and can contribute towards making his run the most impressive . . .'

In five short years, the sonic bang had been transformed from the final trumpet into the aerial drum roll which presaged the greatest show on earth.

Derry's previous appearances at Farnborough had always been in Vampires or Venoms and despite his immense public following, no one had yet seen the first Mach 1 Briton give a public display in an aircraft that was anything like supersonic. Resplendent in their all-white flying overalls, which sharply contrasted with the satin black of their aircraft, Derry and Richards roared off towards Laffan's Plain for their first display on Tuesday. As a Fairey Gannet waddled in to land, the 110's trail suddenly bent sharply as Derry began his supersonic dive, culminating in a searing fly-past at 80ft and spectacular aerobatics. The eagerly-awaited sonic booms were not heard, but as the week went on Derry got his timing down to a fine art such that his 700mph fly-past coincided with the arrival of the bangs.

On Friday 5 September, Derry met Russell Bannock, director of military sales for de Havilland, Canada, near the test pilots' tent. After some chatting, John mentioned that he would soon be looking for pastures new, whereupon Bannock immediately offered the job of chief test pilot should he decide to emigrate. It was yet one more little factor that could have kept Derry from giving the fullest attention to the flying task in hand.

Saturday 6 September was the fourth anniversary of Derry's first supersonic flight in the DH108. This 'public day' dawned cloudy and chilly but the biggest dampener of all came from the engineer who reported that there was a serious problem with the troublesome starboard engine on WG240. There was nothing for it but for John and Tony to fly up to Hatfield, collect first prototype WG236 and rush back for the afternoon's demonstration.

Derry had told the groundcrew to take the starting gear to Hatfield on a low-loader, and the early afternoon was spent pacing up and down waiting for the starting equipment to arrive. If it was delayed much longer, the 110 would miss its slot and the Farnborough arena would be left to Glosters. Just as Derry was about to call the whole thing off, the low-loader appeared: with just enough time to kick the tyres and light the fires, WG236 taxied out and got airborne.

Derry and Richards left Hatfield at 15.15hrs and a few minutes later were over Farnborough climbing to 40,000ft in brilliant sunshine to begin the supersonic dive. On the ground, show commentator Charles Gardner directed

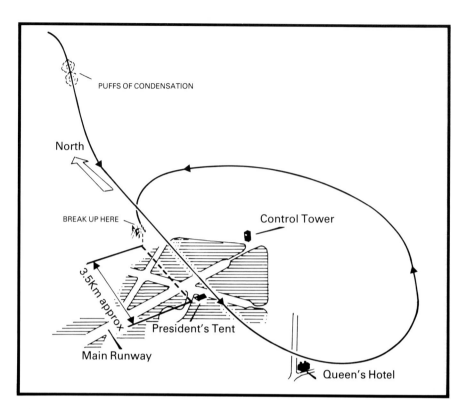

PUFFS OF CONDENSATION

North

BREAK UP HERE

3.5Km approx

Main Runway

President's Tent

Control Tower

Queen's Hotel

Above:
Outline of John Derry's last supersonic dive and display at the 1952 Farnborough Show.

120,000 pairs of eyes towards where they hoped to spy the 110. High above them, John put the aircraft into a dive. The crowd quietened, intent on hearing the sonic boom. The 110 was a bright silver speck against the blue, and as two vapour puffs marked the passing of the speed of sound, moments later there was a shattering triple explosion that shook the buildings. A cheering and clapping crowd saw WG236 pull out sharply and sweep in for a scorching high-speed run from the Laffan's Plain end.

After flashing past at over 700mph, Derry climbed slightly over the famous Black Sheds before banking to the left as usual to begin his long sweeping turn to the north of Farnborough while gaining height and losing speed down to about 450mph. The crowd was quite happy to wait for Derry's reappearance to carry out some aerobatics because the bangs had been the loudest heard that week. As Gardner finished his description of the machine, WG236 swept back into view. With its 51ft wingspan, its 45° swept-back wings and its twin tail booms which curved upwards towards the tail, the large 110 was hard to miss. It was still turning and approaching the airfield around 500ft over Cove Radio Station as Derry prepared to pull into an upward roll. Suddenly, there was a brief scatter of fragments behind the 110 while it was turning and then the whole aircraft reared up violently. A split second later the sky was filled with pieces of broken aircraft. 'Oh my God, this was never meant to happen,'' screamed Gardner, but the warning was too late as thousands of pieces of wing and fuselage arced across the sky. The centre section, booms and fins held together and glided down inverted, but while one engine crashed into wasteland, the other described a graceful arc towards the densely packed Observation Hill. Witnesses said that it seemed to sail quite slowly and gently through the air, straight towards a natural grandstand area where a vast crowd was jammed into immobility by its very size. The one-ton engine fell right among them, making a three-foot crater, and 30 spectators who had come simply to be entertained and view the latest aircraft developments were left dead or dying with another 60 injured.

At the end of the runway in his Hawker Hunter sat Neville Duke. As the last fragments of WG236 were cleared away, Duke opened the throttle, climbed to 40,000ft, and then hurtled down to start his display as the double sonic boom echoed across the airfield. Duke was not being heartless or irreverent towards the memory of a close friend who he acknowledged to be 'the foremost and most knowledgeable pilot in this country in the art and science of high-speed flying'. The show had to go on, and the tradition originating with the Royal Flying Corps that an accident should never be allowed to halt or interfere with a flying programme was adhered to. Derry would have understood, and he would have seen the Duke's display as a characteristic tribute from one great pilot to another. Yet whichever way you looked at it, the 'Thirteenth SBAC Display' had made it clear that however accomplished the pilot, an air display could at any time turn into 'a circus act with no safety net'.

An intense investigation into the crash got underway almost immediately. The investigators' work should have been straightforward because the DH110 had broken up before more witnesses than any other aerial disaster in history. Yet in spite of the comment in the *Spectator* that 'the accident had been published as nothing outside a world war has been published before or since,' there was a far greater measure of agreement on the questions than on the answers. Why, for instance, did the aircraft disintegrate in a moderate turn and not as Derry imposed far greater strain on pulling out of his dive at great speed? There was not even consensus about which bits broke off first, so when a team of Royal Aircraft Establishment (RAE), Farnborough, experts led by Dr Percy Walker, Head of the Structures Department, began their official investigation they had to start with an urgent public appeal for witnesses' accounts and photographs. The response was instantaneous and over 1,000 letters and photographs were submitted by an eager public anxious to help solve the air riddle of the year. Yet there was only one letter which was of some use and fewer than six people who captured to some extent the true picture of events. Most witnesses, in the words of *The Sunday Times*, 'got the split-second time-sequence of disintegration (as finally proved by the research) backwards, filled in bits with imagination, and preferred theories to reports'. Even the experienced Farnborough commentator, Oliver Stewart, was eventually proved to have got it wrong when he declared that the tailplane had broken first. It only went to show that the eye can deceive, and that 100,000 pairs of eyes do not necessarily add accuracy to accumulated testimony.

Meanwhile, the wreckage of WG236 had been gathered from a one-and-a-half mile trail across the airfield and was now all piled up inside a wooden-framed, canvas-covered Bessoneau hangar sited on the southside of the aerodrome. Sir Vernon Brown, head of the Accident Investigation Branch, then specially called for Fred Jones, a young man who was establishing himself as an outstanding detective in the complex field of wreckage analysis. The first few days, as always, were taken up with identifying components and by trying to put the jigsaw back together on a three-dimensional scale to recreate the aircraft. Jones knew that this was a brand new aeroplane, there was a lot at stake, and that any early information would be gratefully received by de Havilland.

It was Fred Jones' habit at an early stage in an investigation to spend a few hours alone with any ruin he was examining to get the 'feel' of the situation. This he did with the 110 on the first weekend, and it was during the Saturday afternoon that Jones decided to make a study of the main separation of the starboard wing. Unknown to him at that stage, he had put his finger on the starting point of the whole disintegration and as the days went by, he felt able to say exactly where the whole disintegration had started. The next question to be answered was why.

Every evening, after a day with the wreckage, Jones would sift through page after page of witnesses' statements. He must have read 1,200 in all, and while each described exactly what the witness had seen, the vast majority saw the accident *only after it had started*. The few who did get it right were over near Cove Radio Station and nearly under, or to the right of, the aeroplane as it approached the aerodrome.

One man who was out near the Radio Station was a professional photographer called Gardner who had captured the accident scene on cine film. He had extracted what he thought were the best bits, and had them published in a well-known magazine in addition to making them available to the accident investigators. Yet as Jones examined the huge sheet of cardboard surmounted by rows of still prints from the film, he realised he was in fact only looking at the effects of the accident. The DH110 disintegrating and crashing might make excellent media copy but it was what *caused* the disintegration that interested Jones. Realising that the photographs started too late in his sequence, he asked RAE to acquire the whole film.

Mr Gardner had done an excellent piece of filming — far better than the newsreels — and his efforts showed that as the 110 was banking to the left and heading towards the crowd, the outer half of the starboard wing broke up. A similar portion of the port wing failed in the same way a fraction of a second later. With only the inner parts of the wings left, the 110 reared up so violently that the cockpit section ripped off under the 'g' forces. However, the engines just carried straight on as the airframe tore itself from them: the tailplane broke up as it was hit by debris from the fuselage. The film showed that the whole episode was over in about one second, so it was not surprising that most witnesses were unable to give an accurate account of events.

Having proved from samples taken from around the starboard wing leading edge, where the failure originated, that there was nothing wrong with the material from which the aircraft was built, Jones concluded that they were looking for a particular loading situation that had exploited the local structure of the wing. At the time of the accident, Derry had the aeroplane in a banked turn and would have had to level the wings and raise the nose to fly over the spectators now ahead of him. To do this, he would have eased back on the control column and also moved it to the right. This latter action would have raised the starboard aileron and introduced a 'twisting form of loading through the wing structure'. Pulling back on the stick would also have added some 'g' or upward bending loading in the wing structure. In Jones' words:

'It was the combination of these two loading conditions which brought about disaster. Their effects were brought together at the leading edge of the starboard wing, just outboard of the integral fuel tank. My analysis showed that compression buckling had developed around the leading edge, and the associated complementary tension stress across the buckle had caused the skin to split. It was from this separation that the major detachment of the wing developed. The wing was torn off from front to rear, just like tearing a piece of folded paper, starting from the fold. The aeroplane started to respond to this sudden disturbance, which, coupled with the loadings already present, set off a chain of actions which included detachment of the port wing, engines, tailplane, cabin and other pieces. The two wing detachments were in such rapid succession that they would have appeared, to a witness, to have been near simultaneous. The result of the loss of the outer portions of the swept-back wings would have caused an aerodynamic centre of pressure shift forward, such that the aeroplane then nosed upward. It was the consequent increasing 'g' loading which had then continued the detachments seen by the vast majority of the witnesses. The wings to them appeared to be symmetrical still, so engines, tailplane and cabin featured high in their memories of the disintegration.'

There was only one safe way to prove this theory and on 4 December de Havilland set up the port wing from a 110 in their Hatfield structural test house. The experiment began with a straight load test without any twisting force applied, and the wing withstood this satisfactorily. Then came the crucial test, simulating the twisting loads of a rolling pull-out by means of rods, levers and hydraulic jacks. It was with mixed feelings that Fred Jones and his team saw a compression buckle develop between ribs 9 and 10 to within 2in of where they predicted the failure had originated.

All previous theories about flutter and suction loads went out of the window. As a RAE specialist wrote to Dr Walker on 9 December:

'. . . it appears that the wing failed in a condition representing a rolling pull-out manoeuvre of $4\frac{1}{2}$ 'g' and 8.5° of aileron at 650mph . . . from the test result the firm have estimated the combinations of normal acceleration and aileron angle which would produce failure and they find that the wing has about 64% of the ultimate strength required in the design. The firm ascribe this failure to an erroneous estimate of the allowable stress at that point.'

Yet if Derry's wing had failed suddenly rather than as a result of a gradual build-up of stress, why had the catastrophe not occurred before in the aircraft's

120hr of flying life? Fred Jones could only conclude that WG236 had never previously reached the crucial combination of forces, although it must have closely approached the single point of stress in previous flights. The aircraft had only to do this once before disaster struck, which led on to the salutary conclusion that if another 110 was to fly a repeat of Derry's Saturday profile, an identical disintegration would have been inevitable.

In the final analysis, it was the wing design that was the culprit. At an early stage, Dr Walker had noted the unconventional design of the 110's leading edge section ahead of the main spar. This thick gauge leading edge was known as the 'D' nose, and while it proved satisfactory on both the Vampire and Venom, it was simply not up to the greater stresses induced on the 110. RAE investigators concluded that a forward main spar or forward vertical webbing in the outer wings would have prevented failure.

The final official Service Accident Report found the cause of the crash, 'considered to be established beyond doubt', to be 'a structural failure of the starboard outer wing caused by the combined effect of pull-up acceleration (associated with turning) and the loads produced by upward aileron (appropriate to straightening out from a turn)'. By early 1953, de Havilland had decided to incorporate a front spar web, thicker wing ribs and reinforced interspar stringers between ribs 8 and 11. After further modifications including an all-moving tailplane, cambered leading edge extensions outboard of the wing fences and reduced ventral fin area, second prototype WG240 flew again on 11 June 1954. Notwithstanding, the RAF was to pass over the DH110 in favour of the Javelin, but the Royal Navy picked it up and had it reworked to sea-going standards as the carrier-borne Sea Vixen.

What came out of the accident? Firstly, John Derry was absolved of all blame. The Farnborough Airfield Commandant, Gp Capt S. W. R. Hughes, told the Coroner's inquest that, 'I am quite convinced that the pilot had no warning whatsoever of the impending failure of his aircraft,' and the jury agreed that, 'No blame is to be attached to Mr Derry'. It was nice to know that the personal distractions and pressures on John played no part in the accident.

Secondly, the Service Accident Report made a telling observation: 'this accident emphasises that for future new designs of aircraft structure, the aircraft should not be flown at high speed or subjected to severe manoeuvres until the agreed programmes of essential major strength tests have been completed.'

It was fortunate that aids such as telemetry and advanced computers were soon to take much of the personal risk out of solving the mysteries of flight, because the loss of WG236 showed that the public arena was no place for rivet-popping contests between competing aircraft manufacturers.

Thirdly, scores of aviators disputed the accident findings long after the event. They themselves 'remember quite clearly' seeing several fragments detaching from the tail *before* the aircraft reared up, and therefore they believe that the tail and not the wing must have been the prime cause of the accident. The eye — even the experienced eye — can play tricks, and one of the aircraft accident investigator's hardest tasks is to differentiate between cause and effect. In the words of Fred Jones, 'Assume nothing — just be certain'.

Finally, the loss of WG236 put an end to the kind of sensational flying witnessed at previous SBAC displays. Although many pilots like John Derry demonstrated a meticulous approach to their displays, Gloster chief test-pilot Bill Waterton likened pre-1952 Farnborough to 'a Roman holiday' where relatively untried aircraft were pushed, often beyond their recommended limitations, as pilots were urged by their employers to outdo the opposition. Certainly, while the dangers to pilots were recognised, those to spectators were but dimly appreciated. Demonstration pilots were given a free hand in their performances and were allowed to bring their aircraft close to spectators even when travelling at speeds not far short of the speed of sound. There seemed to be little thought paid to what would happen in the event of a catastrophe.

This state of affairs had to change and in preparation for the 1953 display, Gp Capt Hughes introduced a new set of rigidly enforced rules. Grass-cutting high-speed runs were now forbidden, as was turning within the aerodrome perimeter if it brought an aircraft into a position where it was pointing towards the spectators' enclosure. The spectators in turn were no longer allowed all over the place. They were now confined behind a 'crowd line' on one side of, and parallel to, the runway, allowing aircraft maximum freedom of movement on the other. Display axes was also to be parallel to the runway. When Gp Capt Hughes briefed pilots in 1953, he told them to liken their aircraft to a loaded gun which could go off at any moment. At no point should it be aiming at the spectators so that if there was ever a repetition of the 110's structural collapse, pieces of wreckage would not be fired towards human targets.

There is perhaps another point worth making. The wise pilot knows that fear is an essential tool of his trade because it keeps him alert and focused. He accepts risk as part of every new challenge, he knows he can be tested by the unexpected, but he counts on his experience, concentration and instincts to pull him through. However, a pilot also needs another essential ingredient, luck. For four days John Derry had performed at Farnborough in WG240, and the second prototype differed from the first in having an aerodynamic fence on the outside of the wing leading edge. Thus WG240 must have encountered near similar conditions over the Cove Radio Station area, but this fence, intended to deal with air-flow problems over the wing, provided just the right amount of external leading edge skin stiffening to hold shape and prevent buckling. Ironically, WG236 originally had similar fences but they were removed as part of the test programme. If John Derry's death saved others by throwing new light on the stresses encountered by new breeds of fast aircraft, the tragic loss of the UK's most outstanding test pilot should also be a reminder to lesser mortals that good fortune will not necessarily always smile on them.

Final Approach

On 17 November 1941, Generaloberst Ernst Udet, one-time outstanding fighter pilot with Richtofen's Flying Circus and now General in charge of Luftwaffe Supplies and Procurement, shot himself. Ill and disillusioned by the failure of his Stukas in the Battle of Britain, and bitter at being blamed for the failure to provide enough new aircraft to sustain the advance into Russia, Udet was still accorded a state funeral on Hitler's orders. Goering, whose withdrawal of support had finally pushed Udet over the edge, tried to make public amends by ordering all *Luftwaffe* dignitaries to be present at Berlin's Invaliden Cemetery to pay their last respects.

One of those expressly summoned to join the guard of honour at Udet's lying in state was Germany's fighter ace of aces, Werner Molders. Now General of Fighters, Molders was carrying out an inspection in the Crimea when the telegram arrived. He took-off at once with his adjutant for the long trip back to Berlin in an He111 belonging to a bomber group at Kherson. The pilot was an experienced 'old hand', Oblt Kolbe, who after stopping at Lvov to refuel discovered that Germany lay under a thick layer of cloud. Kolbe was all for waiting until the skies cleared but Molders insisted on continuing the flight. Popular hero he may have been, but Molders felt under pressure to comply with Goering's personal order and arrive at the funeral in time. He may also have especially wanted to pay a personal tribute to Udet.

Over Poland, a drop in oil pressure forced Kolbe to shut down one engine. Recognising that you can only fly against the odds so far, Molders reluctantly agreed to divert to Breslau-Gandau where he was certain to pick up a rail connection to Berlin. Slowly the He111 descended over Silesia through the fog and rain. Kolbe was locked on instruments when, abruptly, the wires of a cable railway leaped into view. He hauled back on the control column but not before a wire strike disabled the remaining Junkers Jumo engine. As Kolbe tried vainly to glide in, the He111 stalled. One wing struck a factory chimney and the aircraft spun in a few hundred feet short of the airfield. Molders, in the big perspex nose of the bomber where he was probably using his fighter pilot's eyes to try and pick out the airfield lights, was killed outright. The pilot and flight engineer were also fatally injured and only the radio operator and adjutant survived.

Eight days after Udet's state funeral, the great and good of the *Luftwaffe* stood once more round an open grave. It is easy with hindsight to say that Molders had enough flying experience to know he was cutting it too fine, and that he should have been man enough to accept rather than ride roughshod over the professional advice of his pilot. But that is easier to say than do, especially when tunnel vision takes over and you just *have* to get to that wedding, funeral or whatever. Pressure on pilots, placed there by higher authority or by themselves, should never become so great that it becomes impossible to say 'No'.

Werner Molders was a natural pilot and skilful leader, two qualities he shared with another eminent aviator born a few years earlier, AM Sir Harry

82

Above:
Werner Molders with Goering in France, 1940. Despite being famous in his own right, Molders would have found it difficult to resist pressure from anyone so politically as well as militarily eminent as the Reichsmarshall. *IWM*

Broadhurst. A Hendon display pilot before the war, 'Broady' won a DSO and bar and DFC and bar in Fighter Command, going on to become the youngest air vice-marshal when he was given command of that most individualistic of all flying organisations, the Desert Air Force. With the introduction in the mid-1950s of shiny new four-engined jet bombers into the RAF's inventory, a new broom was needed to sweep away some of the mental cobwebs in Bomber Command. Broadhurst, the ex-fighter man, was seen as ideally suited to the job, and in January 1956 he found himself appointed C-in-C Bomber Command at the age of 50.

The mainstay of RAF Bomber Command in the nuclear age was to be the Vulcan delta-winged bomber. Built by Avro to meet a specification calling for delivery of a 10,000lb nuclear weapon over a still air range of 3,350 miles above 50,000ft at a cruising speed of Mach 0.873, the Vulcan was the equal to and in many ways the superior of any bomber in the world in 1956. It was such a source of British pride that soon after XA897 became the first Vulcan to be delivered to the RAF at Waddington on 30 July 1956, it was taken away immediately afterwards to complete service acceptance trials at the Aeroplane and Armament Experimental Establishment (A&AEE), Boscombe Down, before going back to Avro's factory near Manchester to be prepared for a trip to the other side of the globe. In so doing it would send a message that Britain still had the capability as well as the will to defend the furthest regions of the Commonwealth.

83

As part of the crew detailed 'to undertake a route proving and survey flight with a Vulcan to Australia and New Zealand', Sir Harry Broadhurst was to fly in the co-pilot's seat on the righthand side of the Vulcan flightdeck. On his left was the aircraft captain, Sqn Ldr Donald Howard. 'Podge' Howard had joined the RAF in the ranks, had been awarded a commission in 1942 after flying training in America, and had won a double DFC as a low-level ground-attacker in the last year of the war. A decade later, after commanding a Canberra squadron, he had been seconded to Avro as Vulcan project pilot to 'grow up with the aircraft'. At the age of 33, Howard was being groomed for big things in the shiny new Vulcan force.

Behind 'Podge' Howard and the C-in-C, down a short ladder and behind the door well, sat three more aircrew officers facing aft. In the middle was the navigator, Sqn Ldr Edward Eames AFC, a 32-year-old who had been in the RAF for 14 years. To his left was the bomb-aimer's position where the dials and cathode ray tube screen of the Vulcan's navigation and bombing radar should have been. Unfortunately, development of Vulcan avionics was a bit behind that of the aircraft themselves and no airborne radar was as yet fitted to XA897. There being no point therefore in carrying a bomb-aimer, that seat was occupied by Flt Lt (Acting Sqn Ldr) James Stroud, a 29-year-old Vulcan pilot who held a Master Green instrument rating just like the captain. Stroud's presence in the crew had been specifically requested by Howard in case another fully-qualified pilot was needed. 'Tasman Flight', XA897's codename for the trip out east, was supported by three Shackletons and a Canberra PR7. The Shackletons carried groundcrew and support equipment, one staying at Aden and the other two going on to Paya Lebar in Malaya. The PR7 was to act as reserve aircraft: if XA897 went unserviceable, Sir Harry would have leaped into the Canberra to carry on to the next official engagement, leaving Howard and Stroud to bring on the Vulcan later.

The rear crew trio was completed by the signaller, Sqn Ldr Albert Gamble, a 35-year-old Londoner. He sat on Eames' right and was responsible not only for communications but also for all the electric systems that kept the Vulcan flying. By the entrance door there was provision for a sixth crew member, and for the duration of 'Tasman Flight' that seat was occupied by Avro's technical service representative, 38-year-old Frederick Bassett from Wilmslow in Cheshire.

As it turned out, there was little need for Bassett's expertise and even less call for the reserve Canberra. Despite the novelty of everything that lay within it, XA897 behaved perfectly. Leaving the UK on 9 September 1956, it set several speed records on the way out to Melbourne to take pride of place in Air Force Commemoration Week. Then it was on to New Zealand before heading back to Britain.

The return leg was planned to stage through Brisbane, Darwin, Singapore, Ceylon, Bahrain and Cyprus, and by the time it got home, XA897 would have covered more than 26,000 miles. In the end it was the Shackletons that could not stand the pace and by the time the Vulcan was back in Singapore on 24 September, it had proved necessary to amend the itinerary and re-route to Aden where the Shackleton sat forlorn rather than to Bahrain. From there Sir Harry proposed to fly directly home to Boscombe Down on 1 October, missing out Cyprus.

Above:
XA897, the first Vulcan to enter operational RAF service, on public display at Singapore prior to its fateful and fatal return to UK.

Before he left Britain three weeks earlier, the C-in-C had been advised that his Vulcan might be brought back into London Airport to gain maximum publicity. London Heathrow had been Fairey Aviation's flight test centre until 1944 when it was requisitioned by the Air Ministry. It started to take shape as a civil airport in 1950 and by October 1954 it truly became London Airport when nearby Northolt was closed to regular airline traffic. Heathrow had two parallel main runways and between them were the terminal buildings, plus passenger and VIP lounges, which had been opened by HM The Queen the previous December. As the jewel of British airports, Heathrow must have seemed the obvious place to climax such an epic flight.

This proposal received greater support once 'Tasman Flight' proved to be a success, but from Singapore Sir Harry argued against coming into Heathrow because fuel, starter crews and a spare braking parachute would all have to be brought in from Boscombe Down. However, back came a message saying that 'despite your protest, Air Ministry direct you land London Airport and have signalled you accordingly'. All facilities, parachute and servicing party were to be laid on at Heathrow for turn round and departure for Boscombe whenever convenient after arrival, but first the Vulcan crew had to be officially greeted back from their mammoth ranger by a high-powered welcoming committee led by the Deputy Chief of Air Staff, AM Sir Geoffrey Tuttle. Among the party were to be senior Avro personnel, Lady Broadhurst and Mrs Howard.

Awaiting XA897's arrival in Aden was a signal outlining the arrival arrangements at London Heathrow:

'You have been cleared to land at London Airport 011008Z Oct (Repeat) 1008Z (11.08hrs local time on 1 October) for reception at the Central Terminal. You are to join above Amber 1 airway reporting at Seaford, Dunsfold and Epsom. You will be given GCA (Ground Controlled Approach) surveillance and talk-down therefore height must be below 15,000ft by Epsom. In the event of diversion, alternative arrangements will be made. Diversion airfields will if necessary be notified through Southern Air Traffic Control Centre. Agree Boscombe Down or Gaydon if possible, with Waddington, Marham, St Mawgan or Kinloss as area alternatives. Keep rolling and best of luck.'

On 1 October 1956, Vulcan XA897 callsign 'Mike Papa Quebec Kilo 11', left Khormaksar airfield, Aden, at 02.50hrs GMT. The flight across the Mediterranean and then up through France was uneventful, with Donald Howard and Sir Harry taking turn and turn about at the controls. Somewhere near Malta Howard formated with the only other Vulcan then in RAF service, XA895, which was completing intensive flying trials from Boscombe. He asked about the Boscombe weather and was told that it was not very good but expected to clear by 11.00hrs local.

Over France, Howard was told by HQ Bomber Command that visibility at London Airport was 3,000yd in light rain plus 6 or 7/8ths cloud with a base of 700ft. Descending from 42,000ft, the Vulcan entered cloud on joining Amber 1 at 19,000ft and so Howard went on to instruments. As XA897 crossed Epsom at 8,000ft, the forecast landing weather was passed as 2/8ths cloud at 300ft, 7/8ths cloud at 700ft, main cloud base at 5,000ft, visibility 1,100yd, heavy rain and little wind. It was all caused by a front along the south coast but the weather at Waddington was beautiful. HQ Bomber Command said that the decision to come into London 'is up to you and the VIP on board but normal air traffic is having no difficulty. There is high intensity lighting at London.' Although the weather forecast was a bit 'iffy', Howard, in the secure knowledge that he had ample diversion fuel on board, did what any other pilot would have done in similar circumstances. He decided to try one approach at London and if he did not get in, he would divert to Waddington.

After passing Epsom at 10.58hrs local time, Howard was handed over to 'London Radar' which passed a landing QNH altimeter pressure setting of 1,017mb. Both pilot and co-pilot set this on their altimeters. Their windscreen wipers were trying to cope with the heavy rain but Howard took comfort from the reports of high intensity lighting. He was informed that he would be landing on runway 10 Left, and then told to make a righthand turn and to descend to 1,500ft. Given a touchdown height of 80ft, plus allowances for obstacles in the overshoot area, altimeter error and inertia while the aircraft attitude changed from going down to going up, Howard calculated that the 'break off height' for his approach was 300ft. This meant that no lower than 300ft indicated on his altimeter, if he did not have sufficient visual cues to land when he lifted his eyes from the instruments he was to break off his approach, apply power and climb the aircraft away.

At 1,500ft, XA897 was handed over to 'London Director'. As his aircraft approached 5 miles from touchdown at 11.04hrs, Howard put the undercarriage down and reduced to a circuit speed of 150kt. 'London

Talkdown' then took over and as he approached the glidepath, Howard was told to commence descent at 500ft/min. After 15sec the radar controller passed, 'You are 80ft above the glidepath'. Howard overdid the power reduction such that he went 100ft low but at about 1,200ft he increased power and regained the glidepath. Having settled down, Howard then believed that he remained sensibly on the glidepath thereafter, while having to make just two right course corrections.

The rain was lashing down and despite the windscreen wipers, forward visibility from the Vulcan was practically nil. The pilots had agreed that Sir Harry would look out for the approach lights while Howard remained completely on instruments. As the talkdown controller said, 'One mile on the glidepath', the C-in-C reported, '450ft'. Howard remembered being cleared to land as he reached his break-off height of 300ft. Being still on instruments he said, 'Give me the lights'. Sir Harry replied, 'Lights fine starboard'. Howard looked up for the first time and saw lights slightly to the right. Then the C-in-C said, 'You are too low, pull her up', so Howard complied while opening the throttles. At that moment the Vulcan touched the ground.

XA897 had struck a field of Brussels sprouts at Longford 1,030yd from the touchdown point. It seemed to be no more than a glancing blow and at worst both pilots felt that they might have burst a tyre. Neither man suspected that shortly after touching down, the main wheels ran across a ditch tearing the whole oleo leg off one side and the main wheel bogie off the other. On climbing away in a steep attitude, Howard found that the Vulcan kept trying to roll to the right even though he had the control column hard over to the left. Although the power flying controls were still working, they did not seem to be taking effect.

As the accelerating Olympus engines started to roll XA897 on to its back, Howard shouted, 'Get out, get out; it's had it'. For a few more moments he continued to struggle and then ejected. Sir Harry took the controls instinctively but as soon as he pushed the stick forward and applied port aileron, he realised that the controls were not responding. Seeing the bank increasing to about 75°, he repeated the order to abandon aircraft and then ejected too. Simultaneously the radar controller said, '400yd from the runway, talkdown complete'.

From the roof of the Central Terminal building, the welcoming party and newsmen could only watch in horror as the delta reared upwards out of the murk 'like a giant ray leaping from the ocean'. Two staccato cracks signalled the pilots ejecting about 1,000yd down the runway, and then XA897 followed the course of runway 10 Left banking ever more steeply starboard. When it was nearly halfway along, its nose dropped and then the great aeroplane keeled over with mind-numbing slowness to crash on the righthand side of the far end of the runway. It bounced for a further 100yd, a booming explosion rent the air, and the Vulcan came to rest in a billowing mass of flames and smoke. Sqn Ldrs Stroud, Eames, Gamble and Mr Bassett were killed instantaneously.

Howard landed safely on the grass alongside the runway without injury. Broadhurst was less fortunate: landing on concrete between the GCA caravan and where the Vulcan was burning with terrible finality, he fractured bones in his feet and back. Both men were reunited with their wives and then driven away from probing newsmen to the RAF hospital at Uxbridge.

Above:
Sqn Ldr Donald Howard, wearing a borrowed greatcoat, leaves London Airport with his wife after the Vulcan crash on 1 October 1956. *Press Association*

In addition to investigating the causes of such a tragedy, RAF accident inquiries always consider whether anyone failed in the performance of his duty. Inquiry presidents therefore were never junior in rank or seniority to any officer whose conduct, character or professional reputation may be called into question. Consequently, because AM Broadhurst was on the flightdeck, the Court of Inquiry into the loss of Vulcan B1 XA897 was convened with ACM Sir Donald Hardman as president assisted by C-in-C Fighter Command and the Commandant of the RAF College at Manby.

The inquiry found that the weather forecast passed to Sqn Ldr Howard over Epsom accurately described the subsequent weather experienced, and that the captain was justified in deciding to make an attempt to land at London Airport in the prevailing conditions. Nothing was found to be wrong with XA897, yet it still managed to leave wheel marks over a distance of 80ft in a field. There was no evidence to suggest that any part of the aircraft other than the main landing gears had been in contact with the ground, but when both main bogies had been torn off they collided with the inboard elevators on the trailing edge of the huge delta wing. The starboard inboard elevator severed from its control rod and the port probably followed suit. Under the circumstances, the Inquiry believed that this damage 'would have resulted in the pilot being unable to control the aircraft thereafter'.

So what went so tragically wrong between the captain making the right decision to make an approach in the prevailing circumstances and XA897 bursting into flames with four men still on board? Sqn Ldr Howard was graded an 'exceptional' pilot and he had renewed his Master Green instrument rating, the highest rating category in the RAF, in November 1955. But although he had flown hundreds of GCAs on Canberras, he was an inexperienced Vulcan instrument pilot. In fact, although he had had some practice using the GCA at Manchester airport adjacent to Avro's airfield, Donald Howard had never previously flown a full GCA in a Vulcan!

Perhaps his rustiness showed in setting himself a break off height of 300ft. The obstacle clearance limit for runway 10 Left was 150ft and the touchdown elevation was 80ft. To the combined figure of 230ft should have been added an aircraft altimeter pressure error of 60ft and an instrument error of 30ft, making a correct break off height of 320ft. The Inquiry found that the captain had made an error of judgement in setting himself a break off height 20ft too low and also in going below that height.

However, the captain was under GCA control. Apart from right at the beginning of the radar approach, the talkdown controller at no time warned that the aircraft was below the glidepath. On the contrary, about 7sec before the Vulcan first hit the ground and just when the weather was at its worst, the GCA controller informed the pilot that he was 80ft *above* the glidepath. He did not subsequently advise Howard that he was below it and even after XA897 struck the ground, the talkdown continued as if the approach had been normal. The Inquiry concluded that 'the failure to warn the captain that he was going below the glidepath in the concluding stages of the approach was the principal cause of the accident'. The attention of the Minister of Transport and Civil Aviation was drawn to this conclusion 'for such action as he may wish to take'.

Finding such a fast ball coming in his direction, the Minister of Transport and Civil Aviation immediately arranged for an inquiry into the operation of the Heathrow GCA system to be undertaken by Dr A. G. Touch, Director of Electronic Research and Development at the Ministry of Supply. In his report submitted in the middle of December, Dr Touch concluded that there was no evidence of technical failure or malfunctioning of the GCA equipment. His investigations confirmed that the pilot was not warned by the GCA controller of the Vulcan's proximity to the ground, but despite exhaustive examination of the various possibilities, Dr Touch was unable to establish the reasons with certainty. The controller, who was relatively inexperienced, may have concentrated too much on azimuth at the expense of elevation information, but in his defence the Vulcan was some 15-20kt faster on the approach than most passenger aircraft of the period. Dr Touch also believed that even if a warning had been given in the final 5 or 6sec of the 10sec which, in his opinion, elapsed after the pilot was told he was 80ft above the glidepath, it would have been too late. The Touch report concluded that, 'It is very difficult to pass judgement on this matter, but in view of all the circumstances, I do not think the controller should be blamed'. Some argued that Dr Touch did too much special pleading on behalf of the Ministry responsible for the GCA system. His postulation that the Vulcan made a sudden, extremely steep descent almost exactly at the moment the pilot was told he was 80ft high, was rather coincidental. That the Vulcan then quickly levelled out and brought itself into the characteristic tail-down landing attitude is stretching things a bit far, especially as Dr Touch admitted lack of supporting evidence on this point. However, both the Secretary of State for Air and the Minister of Transport and Civil Aviation endorsed the

Touch approach when they announced to the House of Commons that they felt unable to define the degree of responsibility precisely.

XA897 had less than 74 flying hours on the clock when it crashed, so having been barely 'run-in' the Vulcan's write-off value must be put around £750,000 at 1956 prices. Damage to some private property in the tenancy of Mr J. W. Phillip, namely ¾-acre of Brussels sprouts blasted flat by jet efflux, was estimated at £75. Damage to the ditch on land owned by the Ministry of Transport and Civil Aviation accounted for another £10 12s 6d (£10.62½p).

But aeroplanes and crops could always be replaced: it was the loss of four lives on board XA897 that really rankled within the RAF. Because the Vulcan was designed as a high-altitude bomber, it was felt that there would always be sufficient time in an emergency for the rear crew to make a free-fall exit from the floor hatch: only the pilots were given ejection seats specially designed to clear the lofty Vulcan tail. Yet all this theory presupposed that the pilots retained control for long enough to enable the rear crew to reach the door. In XA897's case, the gravitational forces imposed as the crippled bomber climbed and rolled would have held all rear crew members in their seats, including Fred Bassett who was seated just by the door and only needed to roll forward. The morality of giving ejection seats to only two V-bomber crew members was very much open to debate, but the Secretary of State for Air stated that, 'It would be unjust to the pilot and co-pilot were I not to make it clear that it was their duty to escape from the aircraft when they did'.

Nevertheless the question of the pilot in the back of XA897 would not go away and here someone was very much to blame for provoking speculation that might never have arisen had there not been a clumsy attempt to forestall it. Sir Harry Broadhurst was no passenger on the Vulcan flightdeck. He was a first-class pilot who had completed a full Vulcan conversion course and he did not fail his captain in any way. Moreover, he put no pressure on Howard to go into Heathrow. The C-in-C had originally advised against using the airfield and when the weather forecast was received over Epsom, Sir Harry emphasised to Howard that if he was dissatisfied with the prevailing weather conditions he was to divert. Yet someone must have been afraid of adverse public criticism if it was ever revealed that there was another fully qualified Vulcan pilot with a Master Green rating aboard, so the press release stated that James Stroud was the second navigator. Even his death certificate described him as a navigator, a falsification which caused much unnecessary bitterness amongst his family and suspicion among his colleagues.

As it turned out, a whole bevy of Master Greens on XA897's flightdeck would have added little extra safety margin. Although the GCA at Heathrow was a first-generation system that was much cruder than modern landing aids, the full vindication of Howard and Broadhurst had to await the inquest on 30 January 1957. Despite Howard's lack of instrument experience on the Vulcan, the RAF could not credit that he would have gone so far below his break-off height without realising it or that both he and Sir Harry would have misread their altimeters so wildly. A scientific study of Vulcan altimeter errors was therefore undertaken at Boscombe, and it revealed that the large delta wing area created its own atmospheric pressure error of between 70-130ft close to the ground. Friction within the altimeter could add a further 70ft, making a total possible error of 200ft. Add this to runway 10 Left's elevation of 80ft above sea level and it becomes clear that with 300ft indicated on his altimeter, Howard was already among the weeds. There need have been no sudden and unaccountable descent as postulated by Dr Touch: the Vulcan's gentle 3° descent on the GCA glidepath would soon have swallowed up the tiny safety margin remaining to the pilots of XA897. On all future Vulcan instrument approaches, pilots would be briefed to add an extra pressure error correction factor, but it was a pity that it took the tragedy at London Airport to bring the lesson to everyone's attention in the first place.

The loss of XA897 highlighted the importance of being up to speed on instrument flying skills *before* the foul weather comes. It also proved that it does not pay to put your latest, largely untried aircraft into your latest, barely opened international airport for vainglorious reasons. However, 20 years later one would have thought that there was no longer any need to warn against the perils of venturing below instrument approach minima before becoming visual with the ground, but unfortunately there are still too many instances where crews discover this basic truth the hard way.

On 17 November 1988, a slow-moving storm system was located over northwest Colorado. During the next 24hr it would move into Nebraska, producing a northeasterly flow across the Dakotas. This type of air flow, as it lifted over high ground, was ideally placed to produce low clouds, rain and fog, and from approximately 21.00hrs to 02.00hrs on 18 November the cloud base over western South Dakota was expected to be below 1,000ft.

Scheduled to fly from Ellsworth AFB, South Dakota that evening was a pair of B-1B bombers, Uncle Sam's most sophisticated and expensive weapons

Below:
Vulcan XA897's final course as it was talked down by the radar controller towards Heathrow airport. A mild oscillation about the glidepath is not unusual in such precision approaches. At impact the Vulcan was almost level.

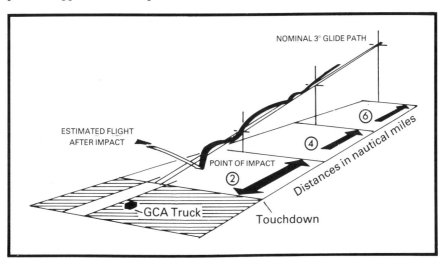

systems. Callsigns 'Kiska 51' and 'Tiger 02', their mission was to fly a low-level route as authorised by the 28 Bombardment Wing Commander. Captaining 'Kiska 51' was Maj Thomas C. Skillman, an apposite surname for an instructor pilot who had flown 505.5hr on the B-1B since May 1987. The remaining three members of his crew comprised another pilot, Capt Mick R. Guthals, an offensive systems officer and a defensive systems officer.

When the crew of 'Kiska 51' arrived at base operations for a weather briefing, they were advised of icing conditions along the designated Instrument Route 32 so that part of the mission profile was changed. By the time replanning was complete, the forecast landing cloud base back at Ellsworth was 500ft scattered, 1,000ft broken, 2,000ft overcast and 5 miles visibility in light snow.

'Kiska 51' took off at 19.29hrs and flew off to do its thing. Back at the ranch flying conditions were supposed to be monitored by the supervisor of flying (SOF). A further fatherly eye was also kept on flying by the 28 Bombardment Wing director of operations. He had approved the cancellation of Instrument Route 32 activity for both 'Kiska 51' and 'Tiger 02', and he was well aware of the forecast weather conditions. However, he did not realise that the weather had changed significantly until 'Poker 94', a KC-135 tanker assigned to the Wing, landed early with a windscreen heating snag. Afterwards, 'Poker 94's' pilot reported breaking out of cloud at 250ft while flying an ILS approach on runway 13. The 28 Bombardment Wing Command Post (Rushmore Control) contacted Ellsworth and asked for a weather update.

At 22.14hrs, Rushmore Control briefed the director of operations on the updated weather situation. Five minutes later he approved an early landing for both 'Kiska 51' and 'Tiger 02', and he sought an assurance that there was an SOF at Ellsworth. Being on the spot, the SOF was supposed to advise both higher authority and aircraft commanders on factors such as changing weather in good time. Rushmore Control told the director that he did have a SOF but this was not strictly correct: instead of being on the ramp, the SOF was in the Bomb Squadron building handing over responsibility to another officer. Thus, when 'Poker 94' landed, neither SOF was advised of its early termination, the weather conditions its pilot reported or that as a result 'Kiska 51' and 'Tiger 02' were returning early. The officers were still completing their SOF handover when 'Kiska 51' crashed at 22.40hrs.

Subsequent investigation of the accident revealed that the B-1B's systems were functioning normally prior to impact. Going over the air traffic transcripts, at 22.19hrs Rushmore Control advised 'Kiska 51' and 'Tiger 02' that the Ellsworth cloud base as a measured 300ft broken, 1,000ft overcast. Rushmore continued by stating that 'Poker 94' broke out at about 200ft when it landed. Both B-1Bs acknowledged this report and their clearances to land early.

'Kiska 51' tried to get in first but at 22.27hrs it overshot from an ILS localiser-only approach when the pilot did not see the runway at the missed approach point. Eight minutes later, 'Tiger 02' landed from a TACAN 31 approach, reporting that he had broken out at the TACAN minimum of 287ft HAT (Height Above Touchdown). This information was passed to 'Kiska 51' and Maj Skillman elected to try a TACAN 31 approach because this promised better results: another partial ILS would have been constrained by a HAT of 467ft. Yet despite his B-1B experience, and all the information from 'Poker 94' and 'Tiger 02', the captain of 'Kiska 51' allowed his aircraft to crash 1,200ft short of runway 31 as he tried to get in for a full-stop landing.

Post-accident analysis of the TACAN approach to runway 31 at Ellsworth revealed a minimum descent altitude error of 20ft on published approach data, which only goes to show that precision approach information is precisely useless if it is not kept accurate and up to date. But the USAF investigation found the main cause of the accident to be that aircraft commander Skillman failed to safely execute an instrument approach. He allowed his aircraft to descend below minimum descent altitude for the approach being flown, which resulted in the B-1B striking fixed ground objects short of the landing runway. Furthermore, the investigation found that the other pilot, Capt Guthals, failed to effectively monitor the instrument approach by not advising the aircraft commander that he was descending below minimum descent altitude without establishing sufficient visual landing criteria.

Yet perhaps it is too easy to ladle all the blame on to the crew of the crashed bomber under the catch-all guise of 'pilot error'. What of the flight monitoring organisation on the ground? It did not come out of the inquiry smelling of roses, and it served as a reminder that a flying supervisor can only supervise effectively if he devotes himself wholeheartedly to being kept in the full picture. But should the buck stop at Ellsworth? Why did a $280+million aeroplane not carry the most basic precision approach capabilities of a fully functioning ILS? Why were the pilots, who were only averaging 20-25 flying hours a month, being put in the position of being expected to fly an expensive asset on a non-precision approach with 200-300ft ceilings at night? Organisations as well as individuals must also learn from flying accidents, and the bigger the organisation the greater the need to stay alert to the possibility of missing the little links in the accident chain.

So even today, flying the most sophisticated aircraft and backed by all the information technology at the disposal of a superpower, pilots can still pay the ultimate penalty if they push, or are pushed, beyond the limits in an effort to get their aircraft down on the ground. There is always a good reason why captains press on — an appointment to keep, a VIP to deliver or simply not wanting to be shown up as a less capable instrument flier than the chap ahead who has just managed to sneak in and land. Perhaps it may dent the image to divert or take guts to stand fast in the cause of safety, but nobody ever said that aircraft captaincy was easy. Just remember the words of the American baseball player Vernon Sanders Low: 'Experience is a hard teacher because she gives the test first, the lesson afterwards.'

Mental and Metal Fatigue

Although the human brain is the most amazing feature of any aircraft cockpit, there can come a time when sheer exhaustion degrades the little grey cells to an unsafe degree. In peacetime it is possible to sustain human efficiency by sensible crew rostering, but in war there may not always be time for such luxuries. Harking back to World War 2, no RAF man was officially allowed on operations beyond the age of 40 and the Medical Branch did not think that young men could stand the strain for long. 'A man subjected to prolonged or repeated fear due to battle stress,' wrote a report, 'will usually persist in fighting that fear as long as his supply of courage lasts. When his courage is exhausted he may either refuse to continue the struggle or develop a psychiatric illness if he has not already suffered death or injury at the hands of the enemy.' For this reason, Bomber Command aircrews were limited to a first tour of 30 operational missions, and thereafter they were sent to recharge their batteries in less demanding roles such as flying training for at least six months.

The combination of fear and fatigue can be a real killer. On the night of 30 March 1944, the might of the British strategic bomber effort was launched at Nuremberg, a city deep in southern Germany which housed one of the MAN heavy engineering works as well as the site of the great Nazi rallies. Beginning with the first take-off at 21.16hrs, a force of 779 bombers eventually joined together over the North Sea in a stream 68 miles long. What they did not know, as they approached their first turning point near Charleroi, was that their radar transmissions had already alerted the Luftwaffe's listening service to what was coming.

Unfortunately the outbound transit not only took place in moonlight without protective cloud cover, but also the bomber crews were briefed to fly in an almost dead straight line for 220 miles across enemy territory before the final turn towards Nuremberg. As ill-luck would have it, this leg took them straight towards radio beacon 'Ida' southeast of Cologne over which the nightfighters of the Luftwaffe's Third Fighter Division were already gathering while their ground controllers tried to identify Bomber Command's objective. The element of surprise, as loitering fighters suddenly found themselves inside a bomber stream, was mutual, but once they had collected their wits the German fighter crews made their kills easily in the moonlight. As heavy bombers started to go down, the defenders sought to capitalise on their good fortune by throwing every available nightfighter into the fray. The Second Fighter Division came down from northern Germany, the First came across from Berlin while the Seventh hurried up from the south to wait for the stream by radio beacon 'Otto'.

It was bad enough that Bomber Command had been routed past 'Ida' and was now heading just to the north of 'Otto', but the fates conspired against the attackers in other ways as well. For a start the winds were much higher than had been forecast, so the stream soon lost its protective cohesion. Then, in addition to the moonlight, the night was so cold that long white condensation trails issued from behind the bombers, pointing them out for all to see. Finally, a *Luftwaffe* unit deposited strings of parachute flares high over the bomber stream to disperse whatever protective cover of darkness still remained.

German nightfighters were therefore able to harry and hunt Bomber Command all the way from the Rhine to Nuremberg. To crown it all, few bombs actually fell on the target from those aircraft that managed to get through the defences. Nuremberg remained shrouded in ground haze, so there was not even the adrenalin of success to raise the bomber crews' spirits. What should have been a routine 'maximum effort' attack had turned into Bomber Command's worst night of the war, and by the time the last stragglers were clear of the skies over German-occupied territory, no less than 94 Lancasters and Halifaxes plus their crews were missing. Weary and dispirited, the remainder started to relax and even descend below oxygen height for a quick bite and warming coffee, but the ill-luck that had dogged this operation from the start had still one more card to play. Northerly winds, which had earlier brought sleet and snow showers to southern Lincolnshire, now spread the bad weather into East Anglia. At the same time, many of the airfields further north were becoming affected by industrial haze and morning mist so that half the bomber airfields were now unusable. It became a race against time to open any airfield in southern England with a runway long enough to take a heavy bomber. As tired and emotionally drained pilots brought their aircraft in, some of which were damaged, some of which were low on fuel, some of which had to

Below:
Following a raid on Genoa in October 1942, Wg Cdr Bruce Bintley of No 102 Squadron landed his Halifax at Holme-on-Spalding Moor, East Yorks, because his home base of Pocklington was completely fogbound. Holme itself was pretty shrouded and before the Halifax could clear the runway, a Lancaster also returning from Italy landed on top of it, killing Bintley and his wireless operator. It was just one of the many tragic accidents that resulted from the strain of continuous operations, and it should be a potent reminder to modern air leaders that they must spend as much effort getting their crews home safely as they do to get them to the target.

be got down very quickly because they had critically wounded men on board, and not a few of which were affected by all three, it was not surprising that the approach to a strange airfield was the last straw. The first crash came at 05.03hrs when Flt Sgt E. R. Thomas' Lancaster flew into the ground at Welford as he tried to get down at the American airfield near Newbury. The No 101 Squadron crew of eight all died. Another 13 bombers followed suit by crashing or crash-landing around England and too many men perished because, after all they had been through, they had insufficient mental or physical reserves left to clear the final hurdle.

In parallel with human fatigue, aircraft components both large and small can also be overstretched to destruction. The strains of old age, or a major change in operating regime for which the aircraft was never designed, are the two classic reasons for structural failure. Combine the two and the risks are greatly increased.

From late 1950 the Handley Page Hastings replaced the Avro York as the RAF's strategic troop carrier. Within three years, fatigue problems with the Hastings' elevator outer hinge-bolts came to light when TG602 crash-landed at Fayid in January 1953. Periodic modification action seemed to keep this trouble at bay over the next 10 years, and if there was any fatigue concern it was concentrated on the wing lower spar-booms which had to be replaced every 4,000hr.

Despite re-equipment with Beverleys and Britannias, by the mid-1960s there were still over 90 Hastings left in the RAF. A proportion were in the Near and Far East, but three home-based Hastings squadrons were concentrated at Colerne to support the strategic reserve, worldwide reinforcements and various forms of training.

On Monday 5 July 1965, Flt Lt Akin and his No 36 Squadron crew, comprising co-pilot, navigator, signaller, flight engineer and air quartermaster, were detailed to fly Hastings C1 TG577 from Colerne to Abingdon to pre-position for parachute dropping sorties on behalf of the Parachute Training School. TG577 arrived at Abingdon that evening, and the following day one sortie was flown in the morning. After the flight engineer had carried out the turn-round inspection, second, third and final sorties were planned to be completed during the afternoon without shutting down the engines. Came the final sortie and TG577 taxied out to Runway 26 with Flt Lt Akin, five crew members and 35 parachute instructors, despatchers and parachutists on board. The weather at the time was good with no turbulence and the Hastings took-off at 16.05hrs, only for the crew to report 'trim troubles' a minute later. This was followed by a transmission referring to 'sloppy controls' and a request for a priority landing on Runway 36. Air traffic control refused because 36 was blocked by a compass-swinging Beverley, but the Hastings was cleared as No 1 to land on Runway 26. The crew acknowledged and announced that they would make a wide circuit and long approach. When well downwind and southeast of Abingdon airfield, the Hastings suddenly reared vertically upwards. It climbed from 1,200ft to about 2,000ft, stall-turned to port, completed about a half-turn of a spin before plunging vertically down into the ground at 16.09hrs. The above average Hastings captain and another 40 persons on board were all killed on impact.

The investigators found the root cause of the accident to be fatigue failure of the two upper bolts which attached the starboard outboard elevator outrigger to the tailplane drag member. Once these two bolts failed, the outer half of the starboard elevator became free and unstable such that the pilot experienced trim troubles and sloppy controls. It was the Board of Inquiry's opinion that abnormal friction in bearings due to distortion of the hinge line, coupled with drooping of the outer half of the elevator, produced the abnormal pitch-up trim resulting in the near vertical climb, stall turn, half-turn spin and plunge into the ground. The Board appreciated that vibration had affected the tailplanes of all Hastings since they had entered service, but this condition had most probably been aggravated by an increased emphasis on low-level operations in recent years. The new low-level operating regime, combined with the demands of continuation and conversion training, was very different from the Hastings' previous and less stressful round of transport route flying.

Because the accident to TG577 was caused by technical failure, no person was found to be blameworthy. The pilot could have done nothing about the situation and it is perhaps for the best that he never realised the seriousness of his position. Had he done so, he would almost certainly have attempted to land on Runway 36 and the Hastings might then have come down on the village of Drayton or the town of Abingdon.

The whole Hastings force was grounded while the tailplanes, elevators, fins and rudders from all aircraft were modified and repaired as required at an estimated total cost of £250,000. Individual Hastings returned to flying duties in dribs and drabs but the loss of TG577 hastened the replacement of the type in the air transport role by the C-130 Hercules. On a wider note, the Abingdon accident made it clear that the structural strength of all aircraft should be checked regularly to reassure crews that the design assumptions, criteria and calculations of yesteryear were still valid. As the RAF's Director of Flight Safety wrote on 29 July 1965:

'I am only too well aware that aircrew and engineers, in order to achieve flying hours, accept unserviceabilities in old aircraft which they would reject instantly in new. We must make sure that this acceptance of the second rate, in an aircraft whose "faults are well-known", does not allow potentially disastrous situations to develop. It must be recognised that there exists an element of, "they have not fallen to pieces for 15 years why should they do so now?". We must recognise that this situation exists and gets steadily worse as aircraft get older. We must take positive action — by survey, recalculation, reconditioning, refurbishing or whatever — to do a little more than is strictly necessary in order to provide an adequate counter.

'I cannot, as Director of Flight Safety, live with equanimity in a situation where we discover the deficiencies of our older aircraft from a series of catastrophic accidents. I appreciate that restudying old aircraft may be expensive. I do not, however, wish to be wise after the event. Such expense is necessary and one which we must recognise is forced upon us whenever we economise by not obtaining younger aircraft of more modern design.'

But even within relatively new airframes, components could fail with equally dramatic results. On 20 December 1950, in the same year as the type entered service, Hastings C1 TG574 was returning from a round trip to Singapore with a flight crew of six and 28 passengers all of whom wanted to get home for Christmas. TG574 took-off from El Adem, Cyrenaica, at 17.58hrs en route for Castel Benito. The night 'was fine with bright moonlight and no cloud or turbulence. Once the Hastings was settled at 8,500ft and 185kt, the co-pilot, Flt Lt Bennett, retired to the ward-room rest bunk and his place up front was taken by Sqn Ldr Jones, a checking pilot.

Then at about 18.45hrs there was a loud report accompanied by a shudder which was felt throughout the aircraft. One of the port inner propeller blades had fractured about 12in from its root. Almost immediately afterwards, severe vibration wrenched the port inner engine from its mounting. Once he had recovered his presence of mind the captain, Flt Lt G. Tunnadine, found that he had lost elevator and rudder control as well as a quarter of his power. He ordered the air quartermaster to move passengers and baggage to the rear of the passenger compartment to trim the aircraft. It was then found that the errant propeller blade had entered the crew compartment through the port side of the fuselage. It smashed through the ward-room, injuring the sleeping co-pilot before coming to rest embedded in the crew toilet on the starboard side. A medical officer among the passengers, Sqn Ldr Brown, came forward to render first aid while distress messages were sent out. Contact was made with Benina airfield and a flarepath was laid out for the Hastings.

Controlling the aircraft by aileron and power alone, Flt Lt Tunnadine spent some time assessing its handling qualities before attempting an approach to land. On being advised by Benina air traffic control that there was a clear run-in from the coast 10 miles away, Tunnadine began his approach from over the sea. He came in at 140kt but shortly before reaching the airfield, he lost sight of the runway lights and touched down some 700yd short of the runway, with wheels and flaps retracted, on gently rising sandy ground with rocky outcrops. At first impact the three remaining propellers, all at high power, were torn off and their engine mountings started to break up. The aircraft became airborne again, scattering debris from the engine nacelles. It touched down 100yd further on with the starboard wing low: this wing was then torn off together with the starboard tailplane. At this point the engines finally detached from their mountings. The Hastings lifted off once more, completing a roll to starboard before striking the ground inverted 350yd down the wreckage trail. It slid a further 70yd on its back, swinging through 180° before coming to rest.

Yet although the Hastings' airframe was subjected to some pretty destructive treatment, none of the passengers inside was seriously injured apart from the medical officer who had remained with the co-pilot until the crash. Sqn Ldr Brown recovered in hospital to receive the George Medal but Tunnadine,

Bennett and two other members of the crew including the checking officer were killed.

Subsequent investigations found that the propeller blade failed due to intercrystalline corrosion, and remedial action was taken across the fleet. The Board of Inquiry paid tribute to Flt Lt Tunnadine when it recorded that, 'the pilot of the aircraft made every effort to effect a successful crash landing. The skill and capability which he demonstrated in flying the aircraft with so little control was responsible for saving the lives of all passengers and two members of his crew.' But it must be said that an equally life-saving factor was Transport Command's policy of fitting rearward-facing and reinforced seats in all their passenger aircraft. If you have a choice, a rearward facing passenger seat is always safest.

Below:
The cause of the accident to TG574 at Benina – a cross-section of the propeller blade that failed through corrosion.

Vulcan B2 XM604 was also only four years old when Flt Lt Peter Tait and his crew were tasked to fly it on 30 January 1968. The five-man crew was programmed to carry out a high-level navigation and bombing training profile and they were joined by the Squadron Nav Radar Leader who was to occupy the Vulcan's sixth seat in order to carry out a check ride. After a delay due to unserviceability, XM604 took-off from Cottesmore, Rutland, at 12.30hrs. During the climb-out, the bomb bay temperature increased and became difficult to control. At 37,000ft the bomb bay temperature of 28°C was close to the top limit, so the captain ordered his co-pilot, Flg Off Mike Gillette, to close the engine air valves. As a precaution, Tait decided to abort the sortie and return to base to carry out continuation training.

For the checking officer's benefit, XM604 carried out an internal aids approach using the Vulcan's navigation and bomber radar. This was followed by an overshoot into the ILS pattern, and the crew could look forward to many such instrument circuits while fuel was burnt off down to landing weight. After flying the ILS approach down to 400ft, the co-pilot applied 80% power to overshoot. At about 700ft, Cottesmore air traffic passed a warning of a slowly moving pop-up radar contact ahead. As he could not see the contact, Flt Lt Tait was cleared to turn left; he took control of the Vulcan and started a port turn at the same time as he increased power to 85%. The aircraft by this time was at 800ft and as it began to bank, there was a loud explosion followed by reverberating thumps accompanied by excessively severe vibration which made it impossible to read the instrument panel. Flt Lt Tait tried to level the wings, only to find that no matter how he moved the control column in any direction, it had no effect. As the bank reached 30° to port, Tait ordered his rear crew to abandon the aircraft: this instruction was acknowledged. The captain continued his attempts to control the Vulcan by use of trim and rudder, and he also closed the throttles, but it was all to no avail and so he ordered the co-pilot to eject. After Mike Gillette had left, Tait looked behind to see what had become of the rear crew, but the black-out curtain (which prevented sunlight from degrading the nav radar's screen) stopped him from seeing anything.

As the controls still had no effect, and as his aircraft had by now rolled to about 130-140° such that Tait was hanging in his straps, he decided to eject. Shortly afterwards, XM604 crashed into the yard of Cow Close Farm some 2½ miles south of Cottesmore airfield. The four rear crew members, having no ejection seats, must have been prevented by 'g' forces from baling out manually and they were all killed.

The co-pilot left the bomber at approximately 750ft while it was banked at about 50°. He made a normal parachute descent but because the surface wind was gusting at about 30kt, Mike Gillette fractured his left upperarm on landing. His captain ejected at around 300-400ft and an estimated speed of 160kt. As Tait had tried so hard to control his aircraft that his Vulcan was by then in a steep nosedown attitude, his ejection trajectory should have been downward. He therefore separated from his seat close to the ground and he should have died because, although his parachute streamed, there would not have been enough time for it to deploy. As luck would have it, Tait then passed through a set of power cables 35ft above the ground. The mass of parachute silk caught on the cables, pulling two of them together and causing an electrical short. The cables acted as a brake to further progress, slowing the captain's fall so that he landed in a standing position below the cables. All Pete Tait then had to do was undo his quick release box and walk away.

An intensive search, which eventually covered an area of four square miles, was mounted using RAF Cottesmore airmen and Army personnel equipped with mine detectors. About one mile to the northwest of the crash site, an assortment of small wreckage was found. In among the debris were blades from No 2 engine low pressure turbine, fragments of No 2 engine turbine containment shield, pieces of No 2 engine rear access doors, part of the top wing skin and stringers from the port side of the bomb bay structure. The remaining major components of all four Bristol Siddeley Olympus engines were recovered from the impact crater, apart from the No 2 engine low pressure turbine disc.

There were 15 eyewitnesses to the accident. In essence they all agreed about the aircraft's behaviour and that there was a fire in the vicinity of the front end of the bomb bay adjacent to the engines. Accident investigators from Farnborough found that the shaft which connected the low pressure compressor to the low pressure turbine in No 2 engine had failed in the region of No 4 bearing. This failure would have occurred about the time the captain advanced the throttles to give 85%rpm, allowing the low pressure turbine to overspeed. This in turn led to the turbine blades becoming detached and to the turbine disc separating from the shaft. These detached components must have been flung in all directions with great centrifugal force, and Flt Lt Tait's loss of control was consistent with the flying controls' circuits and runs being severed in the bomb bay. This conclusion was confirmed by an examination of the debris found well away from the impact centre which showed that this part of the Vulcan had suffered heavy pre-crash structural damage. Some of the bomb bay fragments recovered from the impact areas showed signs of pre-crash burning, confirming the evidence of the 15 eyewitnesses. Such a fire would have followed the opening up of the No 2 engine turbine casing after disintegration of the low pressure turbine.

After the loss of XM604, Avro and Bristol Siddeley co-operated to fit a containment shield inboard of the inner engines to safeguard the bomb bay and another outboard of the outer engine to protect the wing. To prove these shields before they were fitted to all Vulcans, engineers tried to make an Olympus engine fail in a disused railway tunnel near Shepton Mallet. As it happened, the engine proved so resilient that it failed to rupture despite being run for long periods at full power. The only way it could be made to fail within a reasonable timescale was to disconnect the oil supply and hacksaw through the bearing gauge, which only went to prove that the majority of the RAF's flying equipment is basically very good. As with most flying accidents, you have to balance the cost of remedial action with the probability of repetition.

Apart from showing that relatively small failures can kill people and destroy whole aircraft, the loss of Vulcan XM604 also showed that a pilot can stay with his crippled aircraft too long. Flt Lt Tait used up eight of his nine lives that day and there will always be others who are loath to admit that they have run out of ideas. That there may not always be the opportunity to consider all the implications in a rational and timely manner is well understood by some in the US Navy.

'I am an A-7 pilot nearing the end of my second Western Pacific Fleet deployment. I was in the tower during a routine launch. One A-7 was still turning and was parked on the starboard side of the ship just in front of the island. I watched as the blue shirts began to remove the chocks and chains.

'The pilot taxied forward, but when the yellow shirt signalled to stop, the A-7 kept moving. At that point the tailhook came down and the air boss called that the A-7 had no brakes. The aircraft picked up speed and just before it reached the port side of the ship, the pilot ejected. The A-7 went over the side and was under water within seconds. The uninjured pilot was rescued from the sea within minutes — a successful ejection.

'It took only eight seconds from when the pilot started taxiing until the ejection. This didn't leave much reaction time. He made his decision to eject long before the incident or had at least thought about a similar situation before manning the aircraft that day. Think about it the next time you go flying.'

Below:
Gloster Javelin F(AW)9 after coming to rest at Tengah, Singapore in 1966, a bit quicker than the crew expected. The undercarriage collapsed because of wheel-hub fatigue failure.

Right:
Fire has been the prime enemy of airmen ever since they first got off the ground. This is all that remained of Vulcan XL385 after the port engines blew up during take-off on 6 April 1967. Forty tons of blazing high octane aviation fuel soon reduced the bomber to little more than ash – fortunately, the crew got out safely.

103

Low Level

On 16 September 1952, Flt Lt M. A. Clancy was detailed as captain of a Hastings C2 transport aircraft, WD492, on a trip from Thule in Greenland to the ice-cap camp of the British North Greenland Expedition. The camp was located at a far from hospitable 78° 07′ N, 38° 10′ W, and the purpose of the flight was to carry out parachute and free-fall supply drops to the Expedition.

WD492 belonged to No 24 Squadron at Topcliffe but it was no stranger to this type of mission, having previously flown from the RAF Flying College at Manby, Lincolnshire, on long-range navigational training sorties over the Canadian Arctic wastes. The Hastings took-off from Thule at 12.45hrs GMT and it was over the Expedition 3¾hr later. Twelve parachute drops were then made from a height of 700ft whereupon the pilot descended to around 50ft to start dropping the free-fall loads. During the second run, the Hastings' port wing struck the ground. The time was 17.00hrs as WD492 crash-landed approximately one mile southwest of the Expedition camp.

In the subsequent investigation, the primary cause of this accident was found to be that the Hastings was allowed to lose height and turn to port whilst at a dangerously low altitude. It did not help that the aircraft had just entered ice-haze, and there was insufficient ground clearance for the captain to make a rapid transition from visual to instrument flight before his port wing struck the ground. It was also possible that his reactions were dulled by slight oxygen starvation as the Hastings was flying over terrain that was 8,000ft above sea level. The Court of Inquiry concluded that in any future operations of this type, a more reliable method of height indication should be evolved and drops limited to a minimum of 100ft when the containers' fall could be cushioned by snow. It

Below:
Hastings WD492 sits forlorn after its unscheduled arrival on the Greenland ice cap, 16 September 1952.

was further recommended that the co-pilot keep a constant check on the aircraft's altitude when the captain's attention was away from his instruments.

Higher authority concurred with the recommendations apart from the increase in free-fall dropping height:

'Provided the weather conditions allow, free dropping should be done down to 50ft, particularly as experience has shown that damage occurs to the equipment dropped even at this altitude. Furthermore, the digging of equipment several feet out of the snow in Greenland at an altitude of 8,000ft is an exhausting task for the members of the Expedition. The lower the altitude of the drop, the less digging there is likely to be.'

Although the Hastings' captain was held to blame for the accident, no disciplinary action was taken against him. 'The nature of this work was experimental and the experience of others who have done Arctic flying was not readily available. After the crash he displayed a high standard of leadership and power of command.' WD492 still lies preserved under many feet of Greenland ice, but its loss highlighted the perils that can face pilots as they try to balance safety with operating close to the ground in unusual conditions to meet the needs of the 'customer'.

Moving on 16 years and to the opposite climatic extreme, on 7 May 1968 an Argosy freighter flew to Gat-el-Afrag, a landing strip 1,880yd long by 140yd wide in the Libyan desert. The Argosy, XR133, and its crew of five were on deployment from the UK and they were scheduled to make two flights to recover troops, vehicles and freight to RAF El Adem 70 miles to the southeast. After that, the deployment exercise would be complete and it would be time to go home.

The first ferry flight went successfully and XR133 returned to Afrag to be loaded with two Land Rovers, one electric starting trolley, six oil dispensers and six Army passengers. At 10.37hrs the aircraft made a normal take-off from runway 14 whereupon the captain requested permission to carry out an undercarriage check over Air Traffic Control. The Argosy turned downwind for a left-hand circuit at approximately 600ft. It then passed round Rotunda Afrag hill, reappearing in a left turn descending gently down the contours of the hill. When almost in line with the runway, the Argosy turned starboard and although it was very low by this time, the descent continued towards the aircraft dispersal. Alongside the dispersal was a wooden structure, surmounted by a 45gal drum which served as a shower bath. The Argosy's starboard wing leading edge collided with the drum approximately 12ft from the wingtip. Immediately afterwards the starboard wing hit the ground, breaking off the starboard outer wing section and aileron assembly. As they and the 45gal drum flew off, coming to rest around 100yd from the original shower position, the Argosy gained height to around 50-100ft while flames could be seen coming from the break in the starboard wing. The freighter, still with starboard bank on, then lost height again. After trailing the remaining section of the starboard wing along the runway for around 100yd, the aircraft cartwheeled, turned over on its back and crashed into the runway nose first. It exploded and disintegrated, scattering the main wreckage over 300yd and killing the 11 occupants.

It was obvious from the witnesses' evidence that this accident was caused by XR133's starboard wing striking a 45gal drum on top of an improvised shower bath. The impact, compounded by the wingtip then striking the ground, broke off the starboard outer plane section including the outer aileron. The aileron centre section soon followed. The pilot was thereby deprived of lateral control with which to recover from the starboard turn, and his Argosy flew into the ground starboard wing first.

The aircraft struck the drum ostensibly during a low-level undercarriage check. The odd 'red light' after take-off, indicating that one or more of the undercarriage legs had not fully completed its retraction sequence, was not unusual in a region where aircraft had to fly through sand clouds. However, the speed and height of the flypast were not consistent with a normal undercarriage check and it is conceivable that the normally cautious pilot allowed his spirit of exuberance at the completion of a successful exercise to tempt him into a spectacular low run. If this was the case, the pilot's first mistake was in not realising that his previous experience of tactical low flying was limited to relatively low speeds. When flying an Argosy tactically at high speed, a much greater physical effort than he was used to would be required to make any immediate lateral control movements. The pilot was probably late seeing the water shower, underestimated the force required to avoid it, and from then on the outcome was inevitable. That particular 45gal drum was positioned only 10ft above the ground, which was cutting it fine in anybody's language.

It is quite simple to fly safely — all you have to do is aim at the ground and miss! The trouble is that, as over Greenland or at Gat-el-Afrag, there is much more margin for error or time for sorting things out if there is a decent chunk of sky beneath. As a rough rule of thumb, the four most useless things to a pilot are yesterday's weather forecast, fuel left in the bowser, the runway behind and the sky above. All of these factors have their time and place but once military aircraft started to operate lower and lower to avoid radar detection, virtually all the sky seemed to be above. Therefore it became increasingly beholden on pilots to use wisely what relatively little ground clearance remained to them.

On the morning of 11 February 1966, a Vulcan crew based at Cottesmore prepared for a training exercise during the afternoon. In war they would have transitted out from UK at high level because jet engine range was better up there, but as they approached the limits of an enemy's early warning radar cover, they would have descended to hug the ground and thereby hope to evade prying electronic eyes during the final target run. To practise for that day, the Cottesmore crew was to fly Vulcan B2 XH536 for an hour at high-level culminating in a descent to join the Welsh stage of the UK low-level route that ran clockwise over the least populated areas of the country. The weather forecast for that particular area was a minimum cloud base of 2,000ft with

Top right:
Argosy XR133 turns starboard to line up with the Gat-el-Afrag runway for an 'undercarriage check' on 7 May 1968 ...

Right:
... but collision with a wooden shower structure ended the flight in catastrophe.

106

Above:
The remains of Albacore X9117 at Bosham near Chichester in April 1944. The Albacore's crew had been briefed to carry out fighter affiliation with another Albacore down to a minimum height of 3,000ft, but the pair were soon seen dogfighting as low as 600ft. Witnesses then saw X9117 enter a very steep right-hand turn which developed into a spiral dive from which the pilot failed to recover before hitting the ground. The crew perished as the aircraft burst into flames, and all because they ignored their briefing and carried out evasive tactics too close to the ground.

patches at 500ft, and visibility of four miles falling to one mile in snow and 50yd in hill fog.

XH536 took-off from Cottesmore at 13.26hrs. The Vulcan was handed over from Mersey to Southern Radar at 13.59hrs by which time it was at FL420, and the aircraft continued with Southern until 14.48hrs when the captain announced that he would shortly be descending to low-level. Two minutes later he throttled back, extended the airbrakes and came down the slope — all 40-odd thousand feet of it — over the English Channel towards Devon.
'Southern Radar, 47 (Vulcan callsign) has now commenced descent, request Wessex and Holyhead pressure settings for the next hour.'
'Roger 47, the Wessex is 997 and Holyhead 999.'

At 15.00hrs, Southern Radar came back on the air:

'47, what is your passing level?'
'47 is just approaching (Flight Level) 80.'
'47 you are now approaching the limit of my cover, I expect to lose you in 10-15 miles. Will advise you when you fade from cover.'

That did not take long because a minute later:

'47 you are now fading from my screen, ceasing service.'
'47 Roger Southern, thank you very much.'

Passing 5,000ft the navigator confirmed that they had 'coasted out' and were clear down to safety height at 1,500ft. Once there, the Vulcan broke clear of the main cloud base with the grey scud of the Bristol Channel showing through some patchy stratus which broke up as the Welsh coast appeared. Throttling back with airbrakes out, the bomber descended to 500ft. The visibility ahead was four miles and with a clear 1,000ft above to the cloud base, the low-level leg was on.

By 15.08hrs XH536 was driving in fast from the coast to the turning point over Resolven in the Vale of Neath. There was no longer any safety over the flat water and rising ground started to edge the Vulcan up towards the overcast. Down below Mr Evans, a commercial traveller in Resolven, stopped at the sound of the low-flying aircraft and through a break in the cloud he saw it turn on a northeasterly heading.

From here on the gap between the ground and the cloud became narrower as the snow-covered hills merged with the hill fog, and the pilot found himself being pressed up against the cloud base as he strove to keep visual with the ground. On approaching the point where a pair of Hunters had given up and gone home half-an-hour earlier, the navigator would have been urgently calling out the heights while the co-pilot, strip map in hand, strained to identify what he could see of the ground.

Over the next hump and the Vulcan dropped down a little to keep away from the cloud base. Then, out of the mist ahead and a bare 10ft below, a black stone wall flashed across its path. The stick came hard back and as the nose came up, over 50 tons of Vulcan streaked across the short gap between the wall and the grey white shoulder of the mountain. It was now 15.10hrs and at Pant-y-ffordd, about 1,000ft below in the Senni Valley, a noise shook the doors of a cottage. Inside, Mrs Price paused in her household chores and thought it a strange sound to be coming from the quarry at Fan Bwlch Chyth even though they were blasting there. Swathed in the mist 450ft almost directly below the 1,980ft trig point on the mountain, the quarry men heard nothing.

To the search and rescue helicopter crew, the long black streak on the white shoulder of Y Gelli marked the end of the search. They had been out among the hills since first light, but it took less than an hour to find what they were looking for — XH536 which had been reported missing on the low-level route by Cottesmore the previous evening. The harsh clamour of the chopper rotor was muted by the silence of the Welsh mountain as the searchers moved carefully along the half-mile stretch of torn metal looking for signs of life. Two members of the St Athan mountain rescue team were deposited by the wreck but their hopes of finding someone alive were dashed by the stark finality of the torn bomber. There was nothing to do but wait for assistance to arrive while the freezing fog clamped down like a shroud around them. To the Vulcan crew lying by the shattered remains of their cockpit near the icy summit, the coming of the helicopter meant nothing.

Converging on the area by road and air, the Board of Inquiry and accident investigation team found the visibility down to a few feet. At first it was difficult for them to assess the trail of wreckage across the hill. Structural components, pieces of equipment, twisted shapes blackened by flash fire and oddly out of place in the rough frosted grass, appeared out of the mist to be vaguely

Below:

The planned low-level route over South Wales for Vulcan XH536 together with the position of witnesses who saw or heard it pass by. Unfortunately, in drifting off to the left in poor visibility, the crew flew into Y Gelli mountain rather than the valley they were expecting.

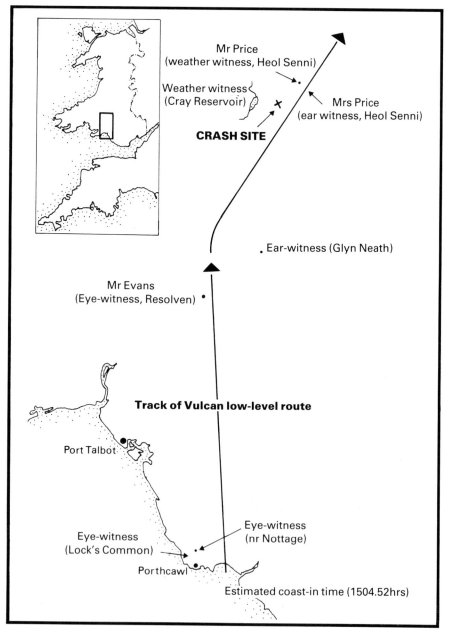

Mr Price
(weather witness, Heol Senni)

Weather witness
(Cray Reservoir)

Mrs Price
(ear witness, Heol Senni)

CRASH SITE

. Ear-witness (Glyn Neath)

Mr Evans
(Eye-witness, Resolven)

Track of Vulcan low-level route

Port Talbot

Eye-witness
(nr Nottage)

Eye-witness
(Lock's Common)

Porthcawl

Estimated coast-in time (1504.52hrs)

identified and then lost again as something else demanded attention. Eventually it became clear that the bomber had struck at a shallow angle on a gently rising slope about 70ft below the 1,980ft summit. Ground markings left by the flying control fairings, tail cone and the deeper incisions of the four jet engines showed that the Vulcan had been practically straight and level on impact. Almost on contact the port wing had ruptured and fuel pumps, shredded tanks and strips of underwing metal had been driven into the soil as the fuel flashed. The fuselage had then slammed down and the nose, undercarriage, underside entrance door and bomb doors had ripped away as the broken aircraft bounced up the hill. Streaming fuel and scattering wreckage, XH536 had then rolled to port and over on to its back. Pronounced markings on the cockpit canopy — it was split almost down the centre — showed that it had been torn off at this point. Here, too, the jet pipes were left behind as the heavy engines tore free to carry on over the summit. The engines had taken a severe hammering but although their compressor rotors were stripped of blades, there were no indications of compressor damage before the crash. Disintegration of the aircraft was almost complete with only the starboard wing outboard of the engines remaining relatively intact. The force of the impact had been such that the ejection seats, plus the rails, had come adrift from the cockpit floor with the pilots still strapped in.

Below:
A frost-covered main undercarriage bogie was just one of the sad pieces of wreckage strewn around the Vulcan XH536 hit Y Gelli in the Brecon mountain range on 11 February 1966.

Once retrieved from the instrument panel, the first pilot's director horizon functioned within specification tolerances apart from a single error which proved to have been caused on impact. Although slightly damaged by fire, the co-pilot's altimeter — reading 2,050ft with a 997mb pressure setting when found — was only to 10ft outside limits at 3,000ft and this was probably due to shock loading. All of which indicated that the Vulcan crashed for no other reason than that the weather conditions at the time were well below the limits for safe aircraft operation in the low-level role. This conclusion was reinforced by the evidence from another Vulcan crew who had been booked into the same stretch of low-level route 2min before XH536. They had taken the precaution of obtaining pre-descent forecasts from two airfields close to the route, and from these the captain considered that it would probably not be possible to get below cloud at low-level. This prophecy was borne out when, as his aircraft approached the Welsh coast around 5,500ft, the captain elected not to descend any further because his Vulcan was flying between layers and the cloud below was unbroken with tops at 4,500ft. The first leg was therefore flown at 5,000ft and during the whole of the subsequent section, the ground was glimpsed only once and then it was a snow-covered hillside which appeared to merge into the clouds.

Following the Brecon Reckoning of XH536, RAF Bomber Command came out with the following statement which is as applicable today as it was then:

'All of us at times bitch about the abundance of regulations that make flying so restrictive. However, we must appreciate that these orders — compiled by experienced operators — are necessary if we are to ensure the safe operation of our aircraft and maintain our strike potential. No pilot has the right to extend himself beyond the limits established by his superiors.'

In other words, obey the rules and maintain situational awareness at low-level such that if the way ahead is no longer clear, there is time to throw it away and climb to the safety of the skies above. Never cut that decision too fine or be tempted into 'press-on-itis' because, as one of Spike Milligan's characters shrewdly observed, 'flying isn't dangerous: *crashing* is dangerous'.

Keep Your Head

A very experienced pilot and navigator were crewed together for a training flight from the Canberra Operational Conversion Unit at Bassingbourn near Cambridge. Part of the conversion required them to take off at night and climb to around 35,000ft to carry out simulated blind bombing using the aircraft's radar equipment. Normal oxygen checks were completed every 10,000ft on the climb, but between 30,000 and 35,000ft, the pilot heard the navigator mumbling to himself on the intercom. He took this to be the normal grumblings of the navigator species when faced with obstinate and contrary kit, but then followed a more worrying period of silence. At one stage the pilot heard his name being called, but thereafter silence reigned and all attempts to communicate with the navigator failed.

In the pilot's words, 'I had to decide whether it was pure intercom failure or anoxia of the navigator. I interpreted the last words as those of a person becoming hypoxic and the final call as a desperate shout from him on realising it.' On making this decision, the pilot began a maximum rate descent to safer levels.

Yet the navigator saw events from a different perspective. During the climb he had been working with his radar equipment and, as the pilot thought, he was having trouble with it. At 34,500ft he noticed that the Canberra had levelled off and then he found that his intercom was not working. He checked his connection but failed to regain contact with his pilot. Almost immediately after this, 'there was a sudden change in the attitude of the aircraft and the altimeter started to unwind rapidly. I could also see lights and clouds through the side window where before there had been only clear sky. This indicated to me that we were in a spiral dive which was rather alarming as I could not talk to the pilot and thought he must be hypoxic.'

The crew were now in the unique position of thinking that one another was hypoxic, which led to some rather interesting results. The navigator bravely left his ejection seat and came forward to try and regain control of the aircraft. This broke his oxygen connection and so the oxygen warning horn began to sound. As far as the pilot was concerned, this noise confirmed his worst fears and he reached forward to switch off the horn. As he did so, the navigator reached the pilot's position and saw what he assumed to be the pilot slumped forward, unconscious, over his control column. To quote the pilot, 'a hand passed my face and started to pull back the control column'. A short struggle then ensued between two men determined to regain control of the aircraft while convinced that the other was suffering from lack of oxygen to the brain. Throughout this altercation the Canberra was descending and by the time it reached 15,000ft, the navigator had operated the pilot's emergency oxygen supply. Believing that he was seeing the pilot recovering, he stopped battling for the control column and returned to his seat where he found that his intercom connection was loose. He quickly re-established friendly relations with his pilot, but two very shaken officers left the Canberra some 15min later after landing at Bassingbourn.

Which only goes to show that under certain circumstances, confusion can arise even between experienced aircrew.

If you are not careful, therefore, some aircraft occurrences can all too quickly take on the aspects of having six barrels of snakes and only four lids. In 1963, a constituted bomber crew plus an umpire were detailed to fly on a RAF Bomber Command exercise. They gathered together in the afternoon for pre-flight planning, and around 18.00hrs they were driven out to their shiny new four-engined jet aircraft that had only left its maker's factory four months earlier to the day.

Engines were started at 19.00hrs and the bomber taxied out normally. It was lined up on the active runway to await the 'Scramble' message which came from Air Traffic Control around 19.15hrs. The dusk take-off – visibility was 3,000yd with a layer of stratus between 800-900ft – was expected to pose no problem for a captain who was an Instrument Rating Examiner.

The bomber lifted off at target speed and the co-pilot selected undercarriage up. Once retraction was complete around 800ft, he selected flaps in. Speed was being held at 180kt and the co-pilot prepared to call 'Climbing away' on the radio. As he reached up to flick the radio selector switch, he saw that the No 2 engine fire warning light was on. He shouted, 'Fire in No 1 engine' and pressed the illuminated warning caption which doubled as the fire extinguisher. The captain looked up, gave No 2 warning light another press for good measure and said, 'It's No 2, not No 1.' The co-pilot completed the immediate fire drills from memory, and then obeyed the captain's instruction to 'Let them know on R/T'. The captain then warned all crew members to check their parachutes.

The first thing the co-pilot noticed after asking Air Traffic for an immediate let-down was that the undercarriage warning flag was flashing on his airspeed indicator. This flag oscillated up and down in the face of the dial if the speed reduced below 160kt while the gear was still up, and the co-pilot noted that the aircraft's speed had dropped to about 140kt. The proper speed for a bomber of that weight was 200kt until the flaps were in, and 240kt thereafter, so the co-pilot gave the control column a gentle push forward and said to his captain, 'Watch your speed'. The controls were sloppy but the captain said, 'No! No! I'm going for maximum height.' At this point a buffeting began which grew until it shook the whole structure. Although the aircraft was around 5,000ft, the captain pulled the control column back again apparently in an effort to maintain the rate of climb.

The vibration continued and the co-pilot shouted out, 'Look out, we are on the judder'. About 15sec later, the bomber flicked over to the left and fell partially inverted into a spin. Although the co-pilot thought that the captain put full right rudder on, he formed the impression that the stick was not pushed forward. The captain immediately ordered his crew to abandon the aircraft, but the only reaction was a voice saying, 'I can't move'. At no time did the co-pilot hear the sound of the cabin door opening, and the 'g' forces only increased as the spin became even steeper once the big bomber completed its first turn. The captain shouted again, 'Get out! Get out!' and when the co-pilot finally managed to focus on the altimeter, he saw that it was passing 2,000ft with the rate of descent needle at full deflection. As three engines were still churning out climb power, the co-pilot reached up to the ejection seat handle without any

114

difficulty and ejected. He landed safely, walked to a nearby road and waved down the first car to pass which drove him back to the airfield from whence he came. The bomber, still in a tight, flat spin, struck the ground with considerable violence two miles from the end of the runway and close to a village. The impact, combined with severe post-crash burning, virtually destroyed the aeroplane. The five remaining crew members died, including the captain who never gave up his attempt to regain control.

The subsequent inquiry found the captain to have been competent to undertake the exercise, the aircraft to have been fully serviceable for flight and the weather suitable. A senior Farnborough investigator examined the aircraft wreckage and found no signs of pre-crash burning in or around Nos 1 and 2 engines. No 2 engine was later sent back to its manufacturer for strip examination but no abnormalities were found. Thus, the Board of Inquiry was forced to conclude that the illumination of the fire warning light had been a spurious indication.

Without the evidence of the co-pilot, the Board would have been hard pressed to establish the cause of the accident. As it was, they still found it difficult to accept that someone of the captain's experience and ability, who had flown a day/night continuation sortie only the day before, could lose control so catastrophically after receiving a single engine fire warning shortly after take-off. The aircraft's pilot's notes state that, 'flight with one engine stopped presents little difficulty and from handling considerations may be indistinguishable from four-engine flying,' so with three engines left turning the captain should have had plenty of safety margins.

Consequently, once they were satisfied that the aircraft did not fail, the Board had to decide why an experienced pilot did. There was no medical evidence to show that he had any disability and the captain was quite calm and reacting normally up to the moment he lost control. To make such a drastic error, the captain must have misinterpreted what was happening and, despite all signs to the contrary, stayed mesmerised by his wrong diagnosis.

The Board postulated that, first, the captain allowed the aircraft to lose speed and stall, and having done that, he did nothing effective about it. We can never know for sure why the captain allowed his speed to fall from 180kt to 140kt when, as the flaps raised, it should have been increasing to a climb speed of 240kt. But at 140kt or 240kt, the large hand – the 'ten' pointer – on his airspeed indicator would have been in the six o'clock position: only the small hand – the 'hundred' pointer would have been different. We must assume that the combined distractions of the fire warning light, the co-pilot's initially incorrect warning, and the need to check that the adjacent engine had no malfunction, took the captain's attention away from controlling his aircraft for too long. On turning back to his flying instruments, he would have expected to see (and thought that he was seeing) 240kt when in fact his airspeed indicator was reading 100kt lower. Such misreadings can occur, but it is not so common for a captain to refuse to acknowledge his error. The co-pilot saw his undercarriage warning flag oscillating but the captain, whose airspeed indicator did not have a warning flag, remained convinced that he was at the right speed. Perhaps, because the co-pilot has misread the fire warning indication, the captain thought he was wrong again.

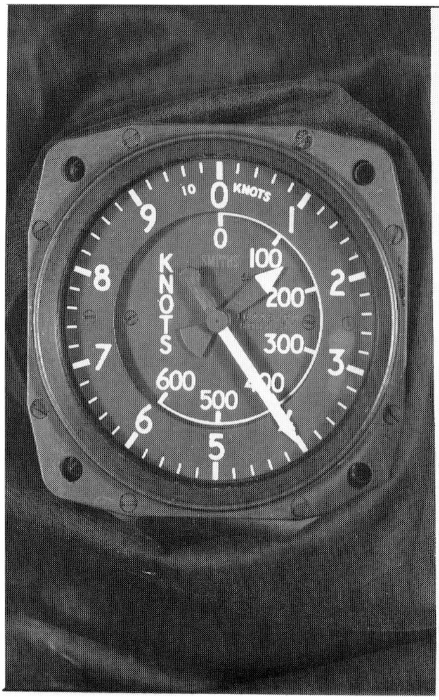
KNOTS
KNOTS

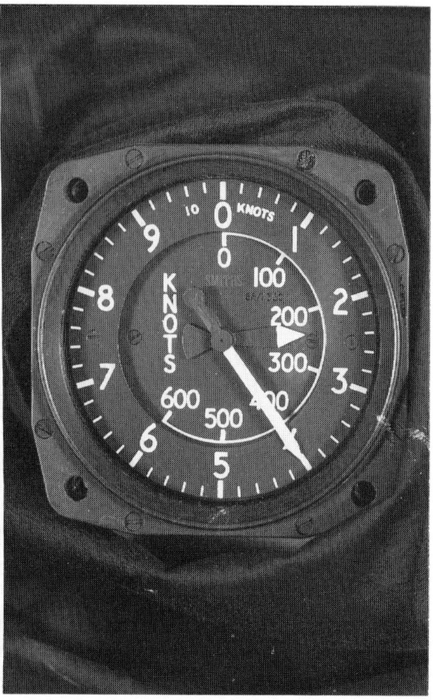
KNOTS
KNOTS

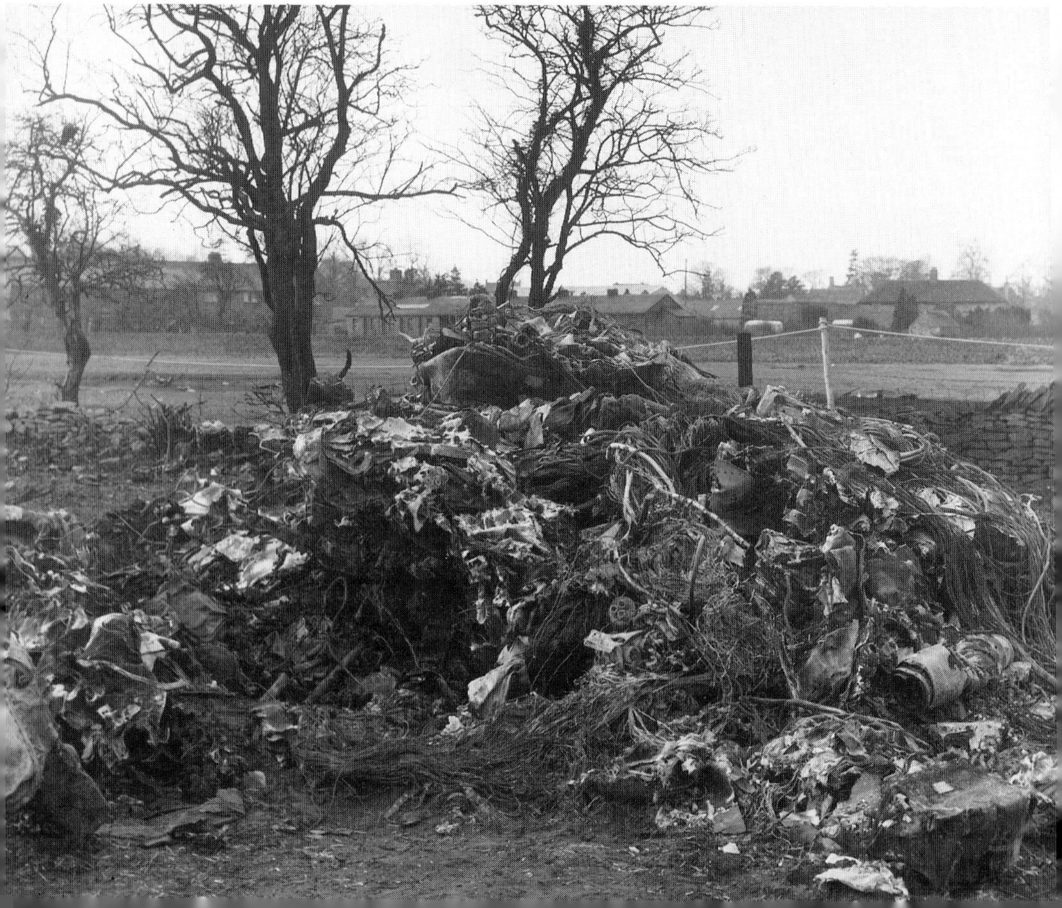

Once the buffeting started, the captain must have assumed that this was an indication of structural failure following the engine fire. Locked into trying to make the apparent facts fit his false reading of the situation, he must have seen it as crucial to strive for maximum height so that the rear crew could abandon the aircraft. The more the aircraft juddered, the more the captain must have thought that the structural failure was getting worse rather than accept that he was witnessing the usual aerodynamic signs of approaching the stall. Once the bomber flicked over, he would not have thought in terms of standard spin recovery.

All of which highlighted the potential pitfalls of distraction. This accident most probably resulted from the flying pilot trying to gain height too quickly because he over-estimated the seriousness of the situation, thereby becoming blind to the crucial need to control his aeroplane. Whatever the degree of distraction during a sudden emergency, if it is not a situation demanding immediate ejection then it is vital to keep the aircraft within the parameters of the safe flight envelope. Even immediate emergency drills must wait until the aircraft is under control: the wing will not burn off in a nanosecond and it will give next of kin little satisfaction to know that their nearest and dearest carried out simulator-perfect checks as they spun into the ground. Along with 'Lookout', the absolutely fundamental rule of safe flying is, FLY THE AIRCRAFT.

Lightning Reflexes

The English Electric Lightning was the first truly supersonic British aeroplane and it soon went on to become the first to achieve twice the speed of sound. Yet despite this great improvement in performance over what had gone before, plus the complexity of its all-weather, radar-interception task, the 'Frightening' was no more difficult to fly than previous RAF fighters. Its abundance of power, coupled with nicely balanced and harmonised controls, made it a natural for display flying. An ability to hurtle from brake release to 30,000ft in two-and-a-half minutes demanded respect, but the Lightning really was simple and viceless to fly so long as its pilot learned about the aircraft and observed the briefed limitations. No one could ever apply to it what an early Lockheed test pilot said of the F-104 Starfighter — 'It won't forgive you a single mistake you make'. The Lightning showed that it was possible to combine safety and 'meanness' in the same machine.

Sqn Ldr Jimmy Dell, permanently seconded to English Electric's flight operations department, delivered the first Lightning to enter RAF service in June 1960. There was much rejoicing in the fighter force on its arrival but once the initial euphoria died down, mutterings began to be heard about drops in serviceability rates, repetitive defects and general weapons system unreliability. Much of this bad name came from the RAF's failure to provide sufficient skilled man-hours to sustain the flying rate, which only proved that there is more to achieving a potent air defence force than simply buying the best aircraft and crews. To keep that force airborne safely, short-sighted economies must not be made when it comes to providing the right level of ground support.

The first dual-controlled version of the Lightning, XL628, made its maiden flight from English Electric's aerodrome at Warton near Preston on the Lancashire coast on 6 May 1959. It had been designated Lightning T4 in time for that year's Farnborough Show, and by the end of September XL628 had completed around 40hr flying time in 92 flights. Flights 93 and 94 were scheduled for 1 October, and Flight 93 went well with XL628 landing free from defects. The second trip of the day was to be flown by Mr J. W. C. 'Johnny' Squier, then aged 39 and English Electric's Chief Production Test Pilot. His task was to measure performance and investigate rolling behaviour at high Mach number, and under the callsign 'Tarnish 2' he took-off from Warton in the prototype Lightning trainer at 10.14hrs before transferring to the control of the Killard Point radar station. Squier, sitting in the lefthand seat, asked to carry out his test run halfway between the Isle of Man and the mainland so that his sonic boom would affect neither. At 10.22hrs the radar controller confirmed that 'Tarnish 2' was 'in the clear' at 'angels 42'.

Right:
English Electric test pilot Johnny Squier, resplendent in pressure jerkin and life jacket, boards Lightning T4 XL628 on 30 September 1959. The next day, the life jacket came into its own as aircraft and pilot parted company over the Irish Sea. *BAe*

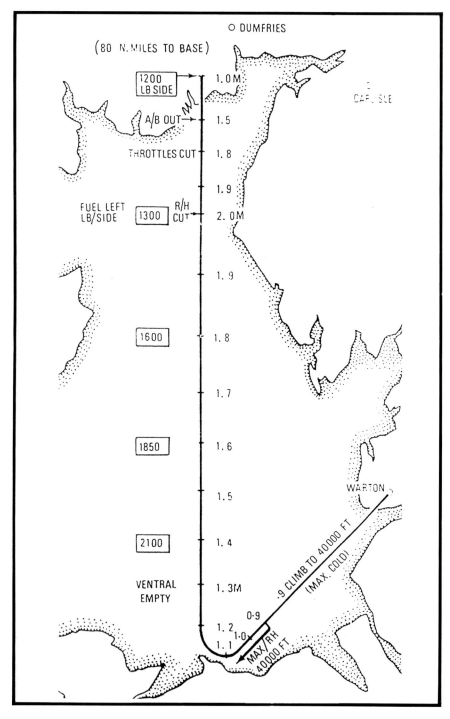

○ DUMFRIES

(80 N. MILES TO BASE)

1200
LB SIDE   → 1.0 M

CARLISLE

A/B OUT → 1.5

THROTTLES CUT 1.8

1.9

FUEL LEFT   R/H
LB/SIDE 1300 CUT → 2.0 M

1.9

1600   1.8

1.7

1850   1.6

1.5   WARTON

1.4   .9 CLIMB TO 40000 FT (MAX. COLD)

2100

VENTRAL   1.3 M
EMPTY

0·9
1.2   1·0
1.1   MAX/RH
40000 FT

Left:
Typical flightpath for a Mach 2 Lightning test flight out of Warton in 1959.

Ten minutes after take-off, Squier was at Mach 1.7 performing a high-rate roll to starboard using full aileron travel. On returning the stick to the centre, a violent yawing motion began. As the side-slip built up, there was a loud 'crump' and the Lightning immediately yawed very violently and flicked over. It then yawed and flicked in the opposite direction. Squier did not have time to get a radio call out and, sensing that he was losing consciousness because the aircraft was gyrating so wildly, he reached up and pulled the face blind handle. He described the one-second delay between pulling the blind and the seat firing as 'the longest second in my life', but when that second was up Johnny Squier became the first man to eject at supersonic speeds anywhere in the world.

Killard Point radar saw the Lightning trace begin to fade at 10.32hrs and when all radar contact was lost seven minutes later, search and rescue action was initiated. XL628 went down in deep water midway between St Bee's Head on the coast of Cumberland and the northern tip of the Isle of Man, but enough wreckage was subsequently recovered to establish that the fin had structurally failed, probably due to the yaw developed by the rigorous test manoeuvre. Lacking the fin's contribution, the Lightning had become stabilised inverted by its anhedral swept wing and was descending almost vertically and very fast when Squier got out. A number of important lessons were learned, not least that a bigger and stronger fin was needed for the T4 and all future Lightning variants.

However, an accident is not over simply when the pilot leaves his doomed ship: he could be injured and not in the most efficient frame of mind but he still has to survive, possibly surrounded by raging seas or inhospitable terrain, until rescue comes. Having got out of the Lightning, Squier kept firm hold of the face blind handle to give a measure of protection because of the speed he was doing. He then started to spin rapidly and the centrifugal force pulled his hands off the blind. His arms became spread-eagled and he feared that he might lose them because the pull was so strong, but then all became calm and serene as there was a change in the axis of rotation. Squier was still sitting strapped into the seat and starting to think that it was time something happened when there was a 'clang' and he felt himself falling away while the seat floated away above him. 'I was not aware of any break in the period, but in actual fact there had been quite a big break as I had been unconscious. The seat separated at 10,000ft and at that point the parachute should have opened automatically, but it did not. I entered cloud and on coming out at the bottom realised that the water was very close. I knew that something had gone wrong and I used the manual override handle to open the parachute. It opened immediately. I thought, "Now for a long wait", and I then hit the water. The cloud base was only 1,000ft.

'I went right down. I had not inflated the life preserver, which I should have done. I inflated it under water, and that pulled me back to the surface. I had to get rid of my parachute, inflate the dinghy and get into it. Having done that, my first idea was to get the Search and Rescue and Homing (Sarah) beacon working. There was a combined microphone and earpiece, but when I put it to my ear, there was nothing from it. I left it switched on and started bailing out

121

the dinghy.' For bailing he initially used his left shoe until he found the proper bailer by accidentally bailing it overboard. He was able to retrieve it because all such 'essentials' were attached by cord to the dinghy.

Squier never did get the Sarah beacon to operate but it did not seem to matter because within an hour or two an American SA-16 amphibian came over at about 300ft and 100yd to his left. He was dismayed to discover that there were only three pyrotechnics in his survival pack, and he was even less pleased to fire one and see it fail to ignite. Later in the day the same amphibian reappeared, again passing to the left. Squier tried to fire another two-star red flare but this time, although there was ignition, its stars failed to fire. When the aircraft was about one quarter of a mile ahead it turned starboard presenting a plan surface. The third two-star red was then fired successfully but by then the amphibian was over the top of Squier and its crew did not see him.

Within minutes of XL628 disappearing off the radar screen, Warton's 'chase' Meteor had been scrambled to carry out a low-level visual search. It flew the same route that Squier had been flying and some time after the US amphibian left, the Meteor pilot dived through the low cloud in almost the right place except that he came down just ahead of the dinghy so that he never saw it. During the afternoon, Squier saw and heard searching Shackletons in the distance. If only he had had more pyrotechnics he felt sure that he could have attracted their attention. He tried to use a heliograph but it proved impossible to direct the sun in the right direction.

Darkness fell and after some difficulty, Squier opened the sea battery of the McMurdo light, put it into the brine to activate the system and was relieved to find something that worked. Seeing no sign of a night search by the rescue authorities, he closed up his dinghy canopy completely and tried to 'fug up' for warmth. It was a very nasty night with rain and 4ft waves, but the dinghy rode the latter comfortably. Nevertheless, Squier's peace of mind was not helped by a fear that his dinghy had developed a leak. Inflatables become softer as the night temperature drops, but the dinghy top-up valve had been fractured during ejection. This meant that Squier could not use the bellows, which was a great worry, and he was unaware at that time that he could top-up by mouth. He dozed intermittently during the night, becoming a bit delirious at times.

With the coming of dawn, the sea calmed and Squier could make out a misty coastline. The shore was some distance away and he tried to paddle towards it with his hands after pulling in the sea anchor. He tried using a knee pad in one hand but that just turned him round in circles. He then came across a piece of driftwood from an orange box, so he was able to paddle just about symmetrically with that and the knee pad.

For a time he made little progress and on occasions he appeared to drift parallel to the coast. Up to then he had not felt thirsty but after paddling for a bit he found around 09.00hrs that he wanted a drink. Unable to locate the desalting kit, Squier erected the dinghy's solar still. He accomplished this without undue difficulty but he was not impressed when he tried to fill up the still and, because of a kinked plastic tube, cold water ran up his sleeve. The first two outputs from the still had to be discarded, the initial one because it was still salty and the second because it had a brackish taste. However, some 4hr after setting up in the desalination business, he extracted about 6oz of water.

After paddling for approximately 8hr, the tide turned and Squier came close enough to part of a wartime Mulberry Harbour to be able to tie up to it. He rested there for a while but then decided that as he was so close in to the shore, it would be better to cast off and try to paddle in for a landing. He cut the dinghy loose using the sharp edge of a tin of Potters Lozenges, and when he was halfway between the Mulberry Harbour and the shore he saw a man on the beach. Squier tried to attract his attention by whistling and shouting but the man's only reaction was to turn and run away. Later Squier learned that this particular beach was private and the dinghy whistle did resemble that carried by policemen.

Below:
Crash and subsequent search area once Johnny Squier parted company with his T4 Lightning. The saga started with the final paints from Boulmer and Killard Point air defence radars as the Lightning fell out of the sky, and ended as both pilot and parts of his aircraft and equipment were washed up in Wigtown Bay.

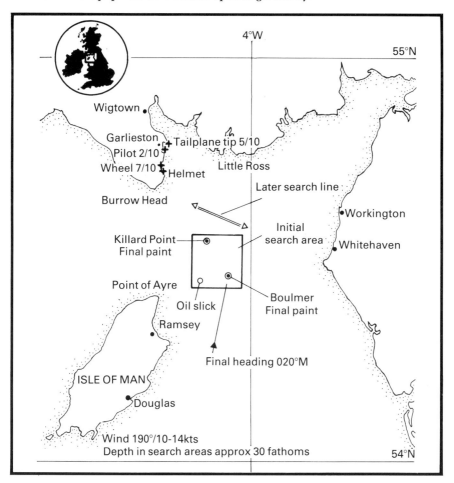

When the dinghy finally beached, Squier stood up. He immediately felt dizzy and it took a 15min struggle to remove his Mae West. After stumbling into the water several times, he eventually found himself on dry ground in Wigtown Bay, western Scotland. After 28¼hr in the water, he was very exhausted and twice tried to climb gates before realising that he could open them. He made for a house, following a path through the trees, and arrived in a garden where he found Miss Donaldson picking roses. She was somewhat startled at his appearance. He said, 'I have just come out of the sea'. She replied, 'Stay there, I have read it in the papers'. Squier was then put to bed before being transferrred to the Garrick Hospital in Stranraer where he was found to be suffering from exposure and crush fractures of the vertebrae. Fortunately he soon regained his sense of humour, telling accident investigators that, 'I wish to say that there is no truth in the rumour that when I recover I am going to spend a holiday boating'. Johnny Squier eventually made a complete recovery.

Twenty-four hours after Squier hit the water, English Electric had given him up for lost. The previous day Warton had sent off its Dove manned by test aircrews to augment the low-level visual search, but this contribution was recalled on the orders of the RAF rescue coordination centre because, as there were already two aircraft searching a very small area in which visibility was poor and deteriorating, the collision risk involved in employing a third aircraft was regarded as unacceptable. Some strong correspondence was to be generated from Warton about this decision, on the grounds that the Dove crew could have done no worse than the 'professionals' who failed to locate Squier's dinghy even though it was floating for many hours within a few miles of the radar-plotted position of the accident. The debate became so heated that it had to be quashed by higher authority, but it tended to obscure the fact that it is far from easy for anyone to locate a one-man dinghy at sea. That was why all aircrews were provided with a Sarah beacon transmitter, and for reasons beyond the RAF's control Squier's became unserviceable. This forced the searchers to resort to visual search techniques, which no matter who had undertaken them, would always have been highly unreliable under the circumstances.

A subsquent check of Sarah beacon batteries in Coastal Command revealed that 20% were below the accepted limits, which highlighted the importance of maintaining crucial survival equipment in tip-top condition. However, even the best kit can only save lives if it is used properly, and that means getting to know *before* the event where equipment is located, how to find it even in the dark, and the drills to get the best out of it. English Electric had designated Johnny Squier as their Flight Safety Equipment Officer but despite his knowledge and expertise, the odds were stacked against him. Things often do not work as they ought, and any pilot could be injured as Squier was during ejection, but someone less informed would have been much worse placed. The sea is hostile enough without having to start on a voyage of discovery to see what survival aids are contained in the life jacket or dinghy. Aircrew must be prepared not only to eject, but also to survive afterwards, because it is just as possible to drown in Lake Windermere as it is in the middle of the Western Atlantic. The only difference is that the former will seem with hindsight to have been such a waste.

Wise men everywhere, from aircraft designers to air-sea rescuers, learn from the experience of others. For instance, the fin of the Lightning T5 — an operational trainer complete with interceptor avionics and weapons — incorporated lessons from the loss of XL628. A T5 began test flying on 29 March 1962 and directional damping with the new 'squared' fin appeared adequate in normal handling with missiles. The final stage of roll-coupling clearance began in June 1965 and all went well with prototype XM966 until 12 July 1965 when Jimmy Dell, with Graham Elkington as flight test observer, set off to undertake the most severe directional proving flight to date involving rapid rolling at Mach 1.8 with the forward-mounted 2in rocket pack extended.

The T5 climbed out of Warton to be handed over to Killard Point. Dell levelled at 35,000ft where he found acceleration to be good because the temperature was an extremely low minus 72°C. He then rolled into a 3g turn to port, extending the rocket pack and switching his test instrumentation on. Full right stick was then applied while still holding 3g and the Lightning initially rolled smoothly. Roll rate then reduced as expected due to sideslip.

Once a wings level, nose-high attitude was reached, the roll-rate reduced to zero. A loud crack was then heard and the aircraft departed violently to the left. Dell's next recollections were of a bang and a strong draught as the canopy blew off and Elkington ejected. Dell must have blacked out during the rapid deceleration of the aircraft: although he must have started to regain consciousness as the aircraft slowed down on passing through 35,000ft, that was just about the time that his observer decided it might be prudent to opt for a Martin-Baker let-down.

As his brain cleared, Dell realised that the aircraft had become inverted at a shallow nose-down angle and that he seemed to be swinging like a pendulum. From his inverted position he had a panoramic view of North Wales, the Isle of Man, the Lake District and southern Scotland, and as the Lightning oscillated from side to side he could faintly hear his call-sign 'Tarnish Six' coming from Killard Point. He did not find it easy to reach up into the cockpit to press the radio button on the throttle, but finally he managed with difficulty to transmit, 'Tarnish Six, Mayday, Mayday, Mayday'.

It was now time to follow Graham Elkington but it was then that Dell realised that his seat straps, which had been specially tightened at the beginning of the flight, had somehow loosened. This was causing the pendulum effect and it meant that Dell could not reach the ejection seat face blind as it was below shoulder level: in addition he realised that if he did manage to eject, he would inevitably suffer back damage because his spine was not firmly in contact with the seat. 'Nevertheless, I only briefly considered undoing the seat harness for a free drop.'

After repeated efforts and a fleeting thought that his wife Marjorie would not be too pleased when she heard about this, Dell finally managed to reach the seat pan handle and pull it. Following a sharp kick in the pants and some slight tumbling the automatics operated, releasing him from the seat and opening the main parachute with a reassuring crack. 'After a short period of stabilisation I was in a steady descent. It was a nice sunny day with no cloud, and the sea looked calm. I watched the aircraft falling below me, still inverted and oscillating until it hit the sea and vanished very quickly.'

Hanging from his parachute, Jimmy Dell had a strange detached feeling and he remembered thinking that having managed to eject, he would now possibly drown in the Irish Sea. He was not unduly perturbed by the thought and at least he assumed that he must be better off than Elkington who had hardly added to his chances by ejecting at high altitude, high speed and in a very cold environment. Something then made him look up and he was mildly surprised to see Elkington's parachute about 300ft above and 300yd to one side.

'I had to force myself to take an interest in the proceedings and took off my flying helmet to drop it as I approached the sea as an aid in judging my height. Unfortunately, having dropped the helmet we both hit the sea at the same time. As I came to the surface I inflated the Mae West but became entangled in the parachute rigging lines: efforts to disentangle were to no avail and increased my feelings of extreme weariness. My next action was to inflate the dinghy and I managed to crawl aboard with some difficulty festooned with the rigging lines. After a short rest I removed the Sarbe Beacon (a smaller and more efficient successor to Sarah) from my Mae West, rigged the aerial in the dinghy and switched it on.

'I then attempted to cut through the parachute rigging lines with my aircrew knife. This was a tedious business with little success and when I eventually dropped the knife in the bottom of the dinghy I lost interest in that activity. I had to keep fighting off the detached "it's not happening to me" feeling. Then I remember looking at my watch and thinking if I'm picked up within an hour, Marjorie and I might make it to the HQ Fighter Command Summer Ball to which we had been invited. At one stage I looked across and saw Elkington's dinghy with him about 300yd away so he was obviously still alive. About this time a Canberra and a Shackleton circled around us and I made a feeble attempt to wave. Sometime after, I heard the sound of the search and rescue helicopter from behind me but I didn't have the energy to turn around to look. I also became aware of an intermittent bubbling sound and realised that the dinghy was deflating. It was later established that when I dropped the aircrew knife it punctured the dinghy.'

The downwash from the chopper followed by the appearance of the winchman rekindled Dell's thoughts about attending the Summer Ball. He warned the winchman that he had hurt his back but despite a sensitive 'lift', Dell had an uncomfortable time lying flat on the helicopter's bare metal floor whilst they picked up Elkington and were then both ferried to Whitehaven hospital. Despite their relatively short time in the water during summer, both men were found on initial arrival to be suffering, among other things, from hypothermia.

Examination of XM966 on the sea bed, plus recovery of the instrumentation pack, showed that the Lightning's fin had failed under the yaw induced by the extreme manoeuvre. It was virtually a repetition of the earlier T4 loss, which showed that the aerodynamics and fin strength relationship was not as well understood as some thought it had become over the previous six years. A massive programme of rechecking the roll-coupling margins of all new Lightning variants was undertaken, while further modifications were introduced to increase the strength of the 'square' fin.

On a personal note, Jimmy Dell never got to the Ball because both he and Graham Elkington suffered back injuries. Dell's were the more serious and he needed a period of specialist rehabilitation before he recovered his fitness to fly again. Nevertheless, their rescue had been a well co-ordinated operation and the time of 1hr from Mayday call to pick up was extremely good considering that the chopper was based 76 miles away at Valley in Anglesey. The rescue was triggered by an alert radar controller at Killard Point who noticed an overlaying of aircraft paints of successive radar sweeps when there should have been a considerable gap. He then made a radio check and received a Mayday call in response. It had been another close call for Warton but this time the RAF rescue services performed superlatively well.

The history of military flight is peppered with examples of people performing considerable feats of survival, but there is probably none stranger than that which occurred at 33 Maintenance Unit, Lyneham in 1964. The MU was then in the process of preparing Lightning F1 XM135 to serve as a supersonic target for its more advanced brethren in Fighter Command, but on its first test-flight in its new role on 29 June, the standby inverter came on just after getting airborne. The pilot, Flt Lt J. Reynolds from Boscombe Down, reported the incident but no fault was found. XM135 was sent off on another air test and the same thing happened again. Another investigation having failed to find the fault, it was decided to isolate the two parts of the standby inverter and test them in turn. As the fault had appeared just after the brakes had been released for take-off, the test would involve opening up the Avon engines on the brakes, releasing them and then immediately closing the throttles and stopping.

Flt Lt Reynolds was not available to carry out the taxy test so his place was taken by a 40-year-old engineering officer, Wg Cdr Walter 'Taffy' Holden. 'Being an engineer, I'd never flown jet aircraft, although I had had pilot training in light aircraft. But I was only to do a simple ground test which included opening the throttle to simulate a fault.' Reynolds had briefed Holden on the maximum rpm at which he could expect the brakes to hold, and then the Wing Commander carried out a full cockpit check. When he was content, XM135 with its canopy removed to enable test wiring to be installed, its ejection seat safety pins in place and its undercarriage ground locks in position, was towed out to the runway.

It was a lovely hot July day as XM135 was positioned at the 18-end of Lyneham's north/south runway which had been closed for the occasion. Holden ran through the cockpit checks from the pilot's notes whereupon the access ladder was removed and the twin engines started. Three test runs were then made, but as they used up less than 300ft of the runway, Air Traffic switched the 36-end traffic lights back to green so that vehicles could resume crossing just over a mile away from the Lightning. Unfortunately, on the fourth run Wg Cdr Holden inadvertently pushed the throttles too far forward and locked them into reheat. His first thought was that the throttles had jammed and by the time he realised his error, the speed of the lightly-fuelled intercepter was building up at an alarming rate. Certainly it must have seemed pretty alarming to the driver of the fuel bowser then in the process of lumbering across the 36-end of the runway which was now much less than a mile away from the fire-breathing tube hurtling towards it. Holden became aware of this vehicle and also of the proximity of the village of Bradenstoke at the end of a runway that was getting closer every second. With only 2,000ft of runway to go and no means of stopping the aircraft, Holden's only course of action was to take-off. Watching with amazement as he narrowly missed a Comet transport, Lyneham Air Traffickers realised what was happening but, as Holden was not wearing a helmet and his radio was not switched on, there was little that they could do other than warn everybody in the circuit to run for cover. They also alerted the emergency services because it seemed a cast iron certainty that their expertise would be needed before the day was out.

Although XM135 had no canopy, Wg Cdr Holden was strapped into the ejection seat. He was not wearing leg restrainers, which would not have mattered had the seat been live, but Holden could not reach the safety pins to make it so. Unable therefore to eject, the good Wing Commander had no other option but to try and land the Lightning. It was a daunting challenge given that despite going solo on a Harvard many years before, his only previous jet experience was a 60min trip in a Javelin T3 two years previously and then he never handled the controls during landing.

As the Lightning accelerated up to 500mph with the promise of even higher and faster to come, Holden realised that his first priority was to remember how to get out of reheat. Having done that, he then thought about landing. Gingerly, he edged round to the duty 07/25 runway and as 07 was closest, he made three attempts to land in that direction, 'but they were totally uncoordinated and I would certainly have made a mess of it and killed myself, so I tried to land in the opposite direction where there were no villages ahead.' While two Britannias were kept 'holding' just outside the circuit, Holden then tried to line up with runway 25 where the 12kt headwind might enable him to make a better approach.

Concentrating on speed rather than height, and turning a little too steeply too low, Holden caused much consternation to those watching from the ground. Nevertheless, although his first approach failed because he ended up too hot and too high, the second was much more successful and the Lightning touched down at something over 160kt with its nose held high. The tail skid hit the runway hard, damaging the braking chute cables which meant that on release the chute fell away without providing any retardation. Heavy brake application plus an uphill runway gradient brought XM135 to a halt about 300ft from the

Right:
George Aird, a de Havilland test pilot, ejects from Lightning PB1 XG332 on 13(!) September 1962. XG332 was being used for missile development work and Aird, an ex-member of the No 111 Squadron *Black Arrows* aerobatics team, was carrying out a demonstration flight when there was a fire in the aircraft's reheat zone. This fire weakened the tailplane control system to such an extent that it failed when the Lightning was at 100ft on the final approach to Hatfield. Fortunately the aircraft nose pitched up, giving Aird those precious few extra seconds to eject successfully even though his parachute canopy may not have deployed fully before he crashed through a greenhouse roof. Despite breaking both legs and right thigh, Aird recovered to resume his flying career. This accident enabled a passing photographer to capture one of the most evocative air accident photographs ever taken, complete with its mesmerised spectator. *Syndication International*

runway end. A much shaken engineering officer finally climbed out of the cockpit with a total solo Lightning flight time of 12min under his belt. He commented later that his salvation was due to a combination of his limited flying experience and the meagre information he had gleaned from the pilot's notes which he had read prior to boarding XM135. The notes themselves stayed on board the aircraft despite the absence of the canopy, but the fact that both 'pilot' and aircraft survived (XM135 was eventually retired to the Imperial War Museum collection at Duxford) says much for the capacity of man in adversity and the importance of knowing your aircraft and its equipment.

Below:
The advent of the high tech age did not herald the demise of the simple accident – Lightning F1A XM188 was written off when brake failure during taxying led it to collide with a squadron office at Coltishall on 2 June 1968.

Lost Prototype

The Handley Page Victor was stablemate of the Avro Vulcan in the RAF's strategic jet bomber order of battle. Built around a crescent rather than a delta wing, the Victor was the most sophisticated of the three British V-bomber types, which probably explained why it was the last to enter service in late 1957. Despite the novelty of much that lay within it, the Victor Mk 1 had a good safety record. The only major glitch occurred when the second Mk 1 prototype crashed on 14 July 1954 after the whole tailplane came away. This catastrophe was caused because tail flutter cracked bolt holes in the fin, allowing the three bolts securing the tailplane to loosen and sheer in quick succession. The cure was to reduce local stress concentrations and stiffen production fins.

Although the Victor Mk 1 had greater range and ceiling than the first generation Vulcan, an improved Victor Mk 2 to counter Soviet air defence advances was not long in coming. Derived from the Mk 1 by the classic 'stretching' process of bigger engines, more wing area and higher all-up weight, the Mk 2 was a longer-ranging Victor capable of safe operation up to 60,000ft and blessed with much more reliable internal systems. The first contract for Mk 2 Victors excluded provision for a prototype so a Mk 1, XH668, was brought forward on the Handley Page production line at Radlett in Hertfordshire to assume the role. Painted overall in anti-nuclear flash white with no roundels, XH668 made its maiden flight in the hands of the company's Deputy Chief Test Pilot, Johnny Allam, on 20 February 1959.

Below:
Prototype Victor Mk 2, XH668, at Handley Page's Hertfordshire factory on 13 March 1959. One of the Victor's pitot tubes is visible just ahead of the port wingtip. The slender tube projected some 6ft forward to enable it to monitor air pressure clear of interference from the wing. Note also the escape hatches above the pilots' seats in the cabin roof: for a time, there was a fear that these might have come adrift causing emergency decompression. *Quadrant/Flight*

After some 97hr of proving flying at Radlett, XH668 was transferred to A&AEE, Boscombe Down on 17 August 1959 for preview handling trials. Boscombe personnel selected to undertake the trials were Sqn Ldrs R. J. Morgan and G. B. Stockman as captain and co-pilot respectively, Flt Lt L. N. Williams as navigator and Flt Lt R. J. Hannaford as air electronics officer; accompanying them was Bob Williams, Handley Page's Chief Flight Test Observer, who was well versed in the management of Victor 2 systems. Both pilots were experienced on the Victor Mk 1 and had been given a familiarisation flight in XH668 by Johnny Allam before the aircraft went to Boscombe. Each pilot also flew another familiarisation trip with Allam on the evening of 17 August.

The first flight of the trials programme was scheduled for 19 August but a minor unserviceability caused it to be postponed until the following day. The schedule for Wednesday 20 August called for a climb to 52,000ft, an hour's tests including high-speed turns to reach the fringe of wing buffeting and to pass a little beyond it, a rapid descent using airbrakes down to 35,000ft, a further series of tests there and at 10,000ft including more high-speed turns before returning to Boscombe after a trip lasting between 2½-3hr.

XH668 got airborne from runway 06 at 10.35hrs local time. Having cleared with Local Control, the crew changed over to Boscombe Approach. A warning of an aircraft passing ahead was acknowledged by XH668 at 10.38hrs but Boscombe had no further R/T contact with the Victor crew. The Approach Controller watched the bomber climb away on radar out to a range of 18 miles, by which time it was heading approximately 020°.

Although a Boscombe Flying Order stated that prototype aircraft should maintain communications with the ground, XH668 was regarded as a new mark of an existing type going off to do nothing that had not already been done by Handley Page and so it was not considered to be a prototype under the terms of this Order. Because of uncertainty as to the progress of the tests plus their intensive nature, it was also deemed impracticable for the crew to check in at regular intervals. Furthermore, because of weather vagaries and the large distances covered at high speed, pre-flight submission of a precise flight plan was not expected and therefore Boscombe had only a rough idea of when and where the crew intended to operate. Whatever the rights and wrongs of such a loose-leash system, no one was officially keeping an eye on XH668. It was only because Sqn Ldr Morgan had arranged with Boscombe's Performance Division to photograph the Victor's landing run that anyone noticed anything amiss as early as they did. At about 13.10hrs a telephone call was made from the Performance Division to Air Traffic to ask for XH668's estimated time of arrival. After a further call, followed by an element of confusion because nobody on the ground was exactly sure how long Sqn Ldr Morgan intended to stay airborne, 'overdue' action was taken at 15.03hrs. By then, XH668 had long since disappeared from the sky.

To try and find out why, a Board of Inquiry was convened at Boscombe on 21 August. Having nothing to go on, they immediately sought help from the public at large. The disappearance of Britain's latest bomber conjured up all manner of theories, ranging from the tail falling off again to highjacking on behalf of an unfriendly agency along the line of James Bond's *Thunderball*. Many people wrote in to say that they had seen an unfamiliar shape 'flying at a great height and very fast', but most sightings were either in the afternoon or in unlikely places such as 'low flying over Kensington'. There were also the usual weird offerings, such as that from the Parisian gentleman who attributed the loss of XH668 to mysterious, but nonetheless damaging, powers unleashed when the moon rose and set.

The inquiry had more luck when Capt Yendall, Master of the small coaster *Aquiety* bound from the Mersey to the Thames in ballast, heard a BBC radio broadcast that an aircraft was overdue from Bosbombe Down. Earlier that day, when the ship was off St David's Head, Pembrokeshire, the Master and two of his crew had been on the bridge because there were extensive fog patches about. At around 11.40hrs the vessel ran out of the fog and there, about five miles away on her port bow, were the Smalls, a little group of islands with a lighthouse. The Master, mate and helmsman then observed a large column of water and spray about 50ft high and some five miles away. This sighting was followed almost immediately by two sharp reports similar to rifle fire. Capt Yendall passed his sighting to Ilfracombe Radio but at this stage there was no way of knowing if it was connected with the loss of XH668, the precise whereabouts of which practically nothing firm was known.

Fortuitously, radar stations kept films of all responses seen on their screens and the Air Ministry asked its UK radar stations to check their photographic records for the period XH668 was airborne. On 22 August RAF Wartling in Kent found a radar track which ended abruptly at approximately the same position and time as that reported by the Master of the mv *Aquiety*.

Two days later both radar trace and Capt Yendall's statement were in the hands of the Inquiry. The track drawn from the trace was then found to be that of a single aircraft which began at a height, position, direction and time corresponding precisely with that of the Victor when last seen on radar by the Boscombe Approach Controller. Because the track was continuous and the responses were good, even at the extreme range of 240 miles from Wartling, it was possible to be sure that the aircraft was above 35,000ft. The area covered by the track corresponded with where Sqn Ldr Morgan normally carried out this type of testing, and the track ended 17 miles in distance, and 3min in time, from the column of water reported by Capt Yendall. The final radar response from the Victor showed it to be in a turn to port and the corrected position of this last response and the position of the splash were found to be only 10 miles apart. This was sufficient to convince the inquiry that at least a substantial portion of XH668 entered the water at the reported position. Their faith was rewarded on 25 August when the first white fragment of rear Victor radome was found by a schoolboy on White Sands beach, St David's. As more pieces of fibre-glass were brought in, arrangements were made for them to be transferred to the Structures Dept at RAE Farnborough, for further study.

At this stage the inquiry felt 'there was reason to believe' that the flight of XH668 terminated abruptly at 11.30-11.40hrs and it entered the sea near the position given by the Master of the mv *Aquiety*. Having said that, the inquiry had no more firm evidence than when it first convened as to *why* the Victor went down when it did. The weather had been good at the time of the accident, and there were no signs of clear air turbulence or jet streams to trouble the crew whose members were more than capable of carrying out the task they had been set. Although the pilots had not amassed many hours on the aeroplane they were flying, that was not unusual for test crews and Handley Page's Deputy Chief Test Pilot was emphatic that he was completely satisfied with the performance of Sqn Ldrs Morgan and Stockman. While at Boscombe, the aircraft was serviced entirely by Handley Page personnel so there should have been no problems there.

All of which left the inquiry to run through every possible cause of disaster they could think of. They concluded that there was an adequate supply of oxygen on board and as individual crew members were wearing air ventilated suits, anti-g trousers, partial pressure jerkins and helmets, they should have been protected from the physiological effects of flying high. No single fault could cause total loss of electric power on the Victor 2, and there was nothing to point towards structural failure because the entire flight was planned to remain within design limitations. There should have been no fuel tank or battery explosion because the Victor was flying with unpressurised fuel tanks and a low-voltage battery that was not crucial to the maintenance of all electrical power.

Moving on to more dramatic causes, it was possible that the bomber had been sabotaged. However, Boscombe was a very secure airfield patrolled at night by police with dogs, and to wipe a Victor off the screen at one fell swoop would have required precise knowledge about take-off time, duration of flight and the vulnerable parts of the aircraft's structure. The sabotage hypothesis just did not seem to be credible.

The last avenue looked more promising even if it was the most haunting. The plotted track of the missing Victor passed close to the missile range at Aberporth which around that period was testing the Bloodhound surface-to-air missile designed specifically to cope with high-flying jet bombers. No novelist would have dared to use the plot but suppose that a Bloodhound had been test-fired out over the Irish Sea and the Victor was heading in that direction. The Bloodhound suddenly went astray, its controller found that his destruct button would not function, and the latest British air defender collided with the latest British attacker. It must have been with a sigh of relief that the inquiry found that there had been no missile launches from Aberporth on 20 August, nor was there any simultaneous activity in adjacent danger areas where interceptors may have been firing guns or air-to-air missiles.

All of these hypotheses were very interesting but they did not lead to any firm conclusions. Accident investigators recalled the fate of the Avro Tudor, a passenger aircraft that was doomed after the crash of *Star Ariel* between Bermuda and Jamaica on 17 January 1949, remained a mystery because the wreckage was lost at sea. Just as there is little chance of securing a murder conviction without a body, the investigation into the loss of XH668 could progress no further in the absence of a substantial amount of wreckage.

By 16 September a systematic search of beaches in St Bride's Bay amassed 34 pieces of radome, one piece of interior cockpit padding and three pieces of aircrew helmet padding. Farnborough tried to extract as much as they could from these finds. Similar pieces of dummy wreckage were suitably labelled, and once the right combination of wind and tide came along, they were dropped into the splash area. The first batch drifted to Ireland and the second was never seen again, leaving Farnborough to fall back on local sea lore which concluded that the wreckage had drifted from a place roughly west of the Smalls.

Unfortunately these pieces of flotsam were too light to be used on crucial parts of the aircraft's structure. What the inquiry really needed were some of the weighty bits of Victor lying hidden on the sea bed. Without them, the inquiry could not even establish if there had been a disintegration in the air, let alone what may have caused it.

From late afternoon on the fateful 20 August, an extensive air and sea search had been going on which by 1 September had narrowed down to a systematic search of an area 16 miles long by 12 miles wide in a south-southwest direction, centred on the estimated position of the splash noted by *Aquiety* and now marked with a Datum buoy. Some 192sq miles might not seem an overly large area, but it approximated to the major part of Greater London from Tottenham to Streatham and from Woolwich to Ealing. XH668 was felt to be somewhere in that area: it might be in six or 600 pieces, but if it was there it had to be found and raised to the surface to re-establish faith in a crucial pillar of British nuclear deterrence.

The trouble was that St George's Channel was nearly 400ft deep where the splash had been seen. Visibility down there was very limited and the sea bed was scattered with rocks plus more than 40 other wrecks. Nothing daunted, the Royal Navy set up an operations room at HMS *Harrier* near Milford Haven from where ships carried out a co-ordinated sweep of the sea bed with Asdic, the submarine detection device. When suitably high intensity echoes were found, wreckage was retrieved by fishing trawlers specially hired for the purpose. As time went on, the number of trawlers would rise from four to 16, supplemented by a special salvage ship *Twyford* which carried a diving chamber, underwater television and huge grabs capable of lifting 20 tons.

Yet despite the resources allocated, days grew into months and Christmas Day came and went without any success. Then on 5 January 1960, the *Picton Sea Lion*, a trawler fishing out of Milford Haven and independent of the search itself, left her nets down instead of picking them up as she approached the designated search area. When the catch was finally examined, it contained a crumpled piece of shiny corrugated metal bearing only a trace of marine growth. The small piece of metal was hurriedly dispatched to Farnborough where it was identified as the first piece of Victor to come from the sea.

At last the sea had begun to show a bit of co-operation but where, along her 12-mile trawl, had *Picton Sea Lion* scooped up what later turned out to be part of the starboard wing trailing edge? *Twyford* was moved out to Asdic sweep along lanes either side of the line of trawl while other vessels swept in between, but by early March nothing further had been found. Most of the main search area had also been swept by this time, so a high-powered meeting held at HMS *Harrier* came to two possible conclusions. Either the splash seen by *Aquiety* was caused by a comparatively small but heavy piece of Victor falling into the sea, or the position given by *Aquiety* was very much in error.

Just as despondency started to settle in, on 13 March the trawler *Clyne Castle*, searching north of the Smalls, brought up four pieces of bent and twisted metal, seven pieces of fibre-glass and three lengths of plastic-covered wire. By noon the next day, all 14 pieces had been identified as wreckage from XH668. After seven months' effort the trail was getting warmer.

Twyford was called into the area to try and find something large on Asdic but to no avail. The gales that accompanied the vernal equinox then drove most of the fleet back into port but on 23 March the trawler *Forards*, operating further west than usual, recovered a piece of engine and some outer skin of one of the bomb doors. The large and unbuckled piece of bomb door was the first indication that the Victor had begun to break up before hitting the sea.

Dr Percy Walker, Head of the Structures Dept at Farnborough and the man tasked with overall investigation into the accident, authorised the diversion of two trawlers to search the new western area. However, while they concentrated on looking for pieces that must have broken away at altitude before gently falling into the sea, the real search effort swung towards where it looked increasingly likely that the mass of the wreckage had hit the water — the area where *Clyne Castle* had made her initial find. For three days *Twyford's* underwater TV camera tried to pierce the sea bed gloom and guide the grab towards more than mud before the trawlers were brought back in again. Overnight they recovered 42 pieces including more engine parts and a whole ejection seat. By the end of March, 100 pieces — or just over 1% of the Victor by weight — had been recovered.

Every time a piece was lifted from the sea bed, it followed the same route: by sea into Milford Haven, from there by road to the RNAS at Brawdy and thence by air to Farnborough. At Farnborough, a full-size skeleton of the Mk 2 prototype had been erected and was waiting: such was the importance accorded to the whole exercise that the second Victor 1 prototype had been dismantled at Radlett and its major components taken down to Farnborough for comparison. The plan was that XH668 would be more or less rebuilt from its broken remains, and in fitting the jigsaw back together it was hoped to discover why it broke up in the first place.

By dint of meticulous manual plotting of each recovered piece, it was soon possible to pinpoint by Decca navigation equipment the area in which XH668 was most likely to be found. As the total score reached 275 pieces on 7 April, a marker buoy was dropped on the new Search Datum. It was over five miles north of the original Datum and *Aquiety's* estimated position of the splash.

Thursday 7 April coincided with the tide slackening sufficiently to let a diver go down in *Twyford's* observation chamber, and he found between 60 and 70 separate pieces of Victor in the first hour. The longest was about 6ft in length half-buried in the sand, and the smallest a mere 3in by 1in of metal shining like a mirror under the light from the three 1,000W lamps above the observation chamber. Up to now the trawlermen had cast and hauled their nets 1,707 times, often in foul weather and with precious little result. Henceforward the fleet of trawlers, varying between eight and 16 ships plus *Twyford*, would carry out 11,069 hauls. By 21 June, 100 days after *Clyne Castle* located the first piece of Victor wreckage, 18½ tons had been found.

XH668 weighed around 63 tons when it crashed of which 20 tons was fuel. The maximum recoverable weight was optimistically put at 40 tons, which left 21½ tons still to be found, but having recovered half-a-ton a day in June, a definite pattern was emerging. Nearly all the wreckage was concentrated in a relatively small circular area not more than 200yd in diameter, but Farnborough was becoming just as interested in wreckage being recovered from the western outer area. It was there that pieces of both port and starboard wingtip — the former with the pitot tube still attached — were recovered showing remarkably little damage, pointing almost certainly to portions of aircraft having broken away before the main mass hit the water. Had they remained attached to the Victor they would have suffered the same impact damage as all the other wreckage recovered from the main area. It therefore became most important to recover more of these relatively undamaged pieces to determine why and in what sequence they had become detached.

Eventually 592,610 pieces of Victor would be retrieved, dispatched on whatever RAE Canberra, Varsity, Hastings, Beverley, Bristol Freighter, Shackleton or Devon could be spared, and taken to Farnborough to be put together again. Individual pieces ranged from a few ounces up to an engine component weighing 570lb, but no matter what the size every fragment was examined — sometimes microscopically — and its place identified in the Victor skeleton mock-up. By 1 July, many of the larger sections were in position leaving, in the words of *Twyford's* second mate, 'fantastically shaped gaps like a giant 115ft-long grey-white pea pod which had been partially eaten away by an enormous and voracious caterpillar'. After they had served their purpose, the smallest bits were placed in a heap in one corner of the hangar, but one piece of evidence that contributed more than most was the co-pilot's wrist watch. Given the shattered condition of the co-pilot's ejection seat, his watch must have been in the aircraft when it struck the water. The hands on the watch face had stopped at 11hr 30min and 46sec, and experts concluded that at this time the watch had been stopped by a single violent blow. Colleagues of the co-pilot were

Below:
The face of the co-pilot's watch recovered from the wreckage of XH668. Expert opinion was that the watch had been stopped at 11hr 30min and 46sec by a single, violent blow, which was most likely the Victor hitting the sea.

unanimous in stating that he was meticulously precise in setting his watch, but the master of *Aquiety* had seen the splash at 11.40hrs so there was now a discrepancy in time as well as position. *Aquiety's* log was obtained and the reported bearing and estimated distance of the splash were found to have been extremely accurate. It was where the coaster was said to have been at the time of sighting that was wrong. That was her position when the entry was actually written into the log, not where she was when her crew saw the splash. Once the true position of the splash was calculated, it came out almost exactly at the location of the main wreckage. If only the searchers had known that nine months earlier, it would have saved so much time and effort.

The combined evidence from Wartling radar and the co-pilot's watch indicated a time of descent not exceeding 1¼min. The condition of the wreckage was also consistent with a rapid descent and Farnborough was able to show that XH668 had struck the sea at around 700mph whilst diving steeply. Calculations indicated that the prototype would have rapidly gone supersonic at high altitude — the sharp reports heard on *Aquiety* were probably sonic bangs — but the speed would have become subsonic again in the denser atmosphere lower down. Nevertheless, a descent speed around 600kt would have been sufficient to tear XH668's bomb doors and wingtips away between 8,000ft and 5,000ft. Bomb door disruption was not unexpected but it took wind tunnel tests to establish that the wingtips would have fluttered at the appropriate height as

Below:
Geography of the main features of Operation 'Victor Search' following the loss of XH668 on 20 August 1959. The Reported Position of Splash acted as the Search Datum around which the Search Area was set up.

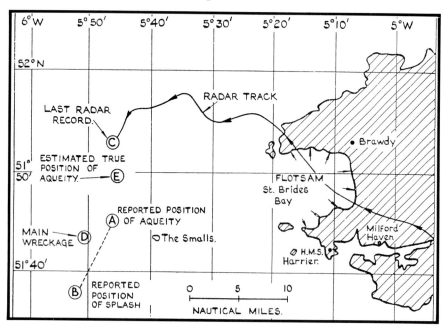

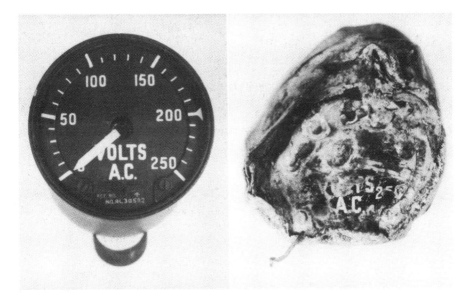

Above:

A voltmeter recovered from XH668 (right) compared to an intact version. Although it might seem to be battered beyond recognition, close examination of the voltmeter revealed that it had been registering 200V up to the moment of impact. This showed that the aircraft's electrical services were functioning satisfactorily.

the speed passed from supersonic to subsonic. Although the wingtips would have disintegrated with the onset of flutter, the remainder of the wing structure stayed outside the flutter range and reasonably intact.

The aeroplane had remained in a clean configuration with flaps, undercarriage, and airbrakes retracted, but rather surprisingly the nose flaps along the outboard wing leading edges were found to have been extended. All four engines had been running at high power with throttles set forward, and there was no evidence of any fire. An interesting find was one of the voltmeters which, although crushed on impact, showed a faint mark on the dial when viewed through a microscopic lens. The needle itself was still under the sea but the faint mark showed where it had been smashed against the face of the dial as the aircraft hit the water. The meter was registering 200V at the time which showed that the Victor's main electrical system was working right up to the end.

Two weeks after XH668 went down, a Victor 1 also flying out of Boscombe had spontaneously lost both pilots' roof hatches due to misalignment of the quick release catches. For a time it was feared that the same might have happened to the Victor 2 prototype, causing the crew to lose consciousness immediately above 50,000ft, but when XH668's hatches were eventually retrieved they were only slightly damaged and appeared to have been intentionally jettisoned below 10,000ft. The captain's ejection seat had left the aircraft in the correct manner but although he had separated from his seat, his parachute would have been unable to deploy properly. The co-pilot's seat had not left the aeroplane but the evidence showed it was still occupied and in the process of ejection. The rear crew seats were barely recognisable.

Twyford eventually had to back off from recovery work because its great grab, originally brought in to lift huge chunks of Victor should they be found, was becoming too crude a retrieval instrument now that the bomber was found to consist of little more than small pieces of debris. More trawlers were brought in during the summer because, despite being a long way down, the sea bed was most satisfactory for trawling. It consisted mainly of sand with a fairly level surface, and although tidal movements and currents gradually buried wreckage that had not been recovered, special rakes were developed which pulled debris out of the sand to be caught in a net trailing behind. All manner of surprises surfaced in this fashion, not least being the bottom half of a lady's long evening dress; many a matelot speculated on the background to its being on the sea bed! More seriously, in mid-September a haul brought up what appeared to be a human rib cage. No bodies of the Victor crew having been found, this object was hastily dispatched to the aviation pathologists at Farnborough only for the pelvis to be found to have come most probably from a two-year-old steer.

But by this time, Dr Percy Walker and his team had a pretty good understanding of what had gone wrong with XH668. Knowing that the pilots were fully conscious and not injured in any way, and having recovered motors, generators and hydraulic pumps that showed signs of running just before final impact, the inquirers had to consider loss of control brought about by other factors. Going back to the beginning, XH668 and its crew had been sent to operate up to a speed of Mach 0.97. As the Victor approached the speed of sound, aileron control would have virtually disappeared because the power flying control jacks would have been unable to supply the heavy forces required to operate the down-going aileron. This would not have been critical when flying straight and level, but the crew was briefed to carry out buffeting steep turns and buffeting would have restricted the amount of extra lift the elevators could generate. The combined effects of all these factors meant that a spiral dive could have developed from which recovery became less and less likely as speed increased.

Yet even if this happened to the Victor, it was the result and not the cause of the accident. This issue rested on what caused the Victor to exceed Mach 0.97 in the first place, and given that a skilled test crew was involved, the inquirers had to find a cause which distracted the pilots in a major way at the critical time.

Right:
The Victor Search Area, divided into 12 squares each 4 miles × 4 miles. Areas 'Jig', 'Saw' and 'Puzzle', being in the centre closest to the reported splash position, were felt to be most likely to contain Victor remains. If only the searchers had known *Aquiety's* true position from the beginning, it would not have taken so long to discover that most of the thousands of pieces that once comprised a Victor lay over the mud and sand in area 'Water'.

The diagram also shows the approximate position of the various sonar contacts (A, B, etc) and the trawl line of *Picton Sea Lion* on 5 January 1960. The dotted line shows the extent of the area swept by trawlers up to 24 February 1960.

Among the recovered wreckage to be re-assembled and examined were the wingtips which came adrift during the descent. Victor 2 wingtips were more or less self-contained and detachable units, and each carried a long, slender tube projecting some 6ft ahead of the wing leading edge comprising both static and pitot pressure systems. When the wreckage of the port wingtip was re-assembled it was noted that the pitot tube was bent through more than 90° at

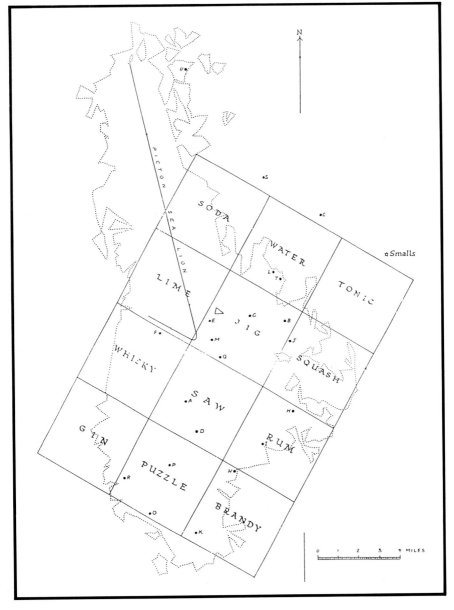

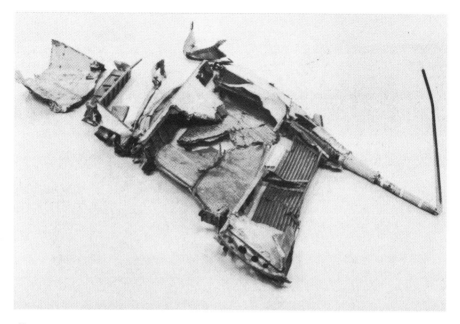

Above:
Wreckage of XH668's reassembled port wing tip showing that despite the strains which bent the pitot tube through 90°, the tube still held in its mounting.

the end of the supporting sleeve that constituted part of the mounting. This damage was probably caused by flutter late in the descent but it was significant that, despite the severe strains imposed, the pitot tube base remained firmly held in its mounting.

The re-assembled starboard wingtip told a different story because, although structural damage was similar to that on the port side, the pitot tube was missing in its entirety. A Victor pitot tube was held in position by a form of chuck in which a conical sleeve was tightened against two tapered collets. There was nothing wrong with the design but the starboard pitot tube had come cleanly and neatly out of its mounting without any signs of damage. Furthermore, when the tapered sleeve for the starboard mounting was recovered from the sea with the collets still in position, the inside of the sleeve was found to be covered in protective paint which had clearly been applied too freely. It was surmised that the sleeve had originally been tightened adequately but only against localised areas of paint. As the paint wore away or became distorted, progressive loosening of the collets' grip upon the pitot tube could reasonably be expected. Nothing similar was found on the port installation.

The main flying instruments were duplicated — the port pitot tube fed the captain's instruments while the starboard fed the co-pilot's and navigator's — and the pitot systems were in many ways identical, but the starboard was unique in serving the 'stall detector' and the Mach trimmer. Wind tunnel tests a decade earlier had predicted that the crescent wing stall would be pretty vicious, so Handley Page developed great accumulators of stored energy which

142

Equivalent picture of XH668's starboard wing tip, showing that on this side the pitot tube had come away from its sleeve mounting.

thumped leading edge flaps down within a second on a signal from a pressure ratio switch which calculated lift coefficient from tappings below the wings. While the flaps travelled, warning lights illuminated in front of both pilots. As it turned out, the stall was nothing like as bad as feared so the nose flaps would eventually be locked up on the Victor 2, but in 1959 XH668 still had them. In addition to pressure ratio switch signals, the stall detector received pitot and static pressures from the starboard pitot tube.

The Mach trimmer was incorporated to counteract the nose-down trim change caused by air compressibility at high Mach number. As the aircraft accelerated from Mach 0.85 to Mach 0.95, the Mach trimmer raised the elevators through an angle of about 5° without altering the stick position, and irrespective of the pilot's control inputs, in response to pressures conveyed from the starboard pitot tube.

With all this information at their disposal, Dr Walker's team set off on a 'what if' exercise. They started by assuming that XH668 was flying buffeting turns around Mach 0.94 at 52,000ft. Suppose then that the starboard pitot pressure tube started to come adrift slowly under the effect of all the buffeting. A leak would slowly develop causing a fall in indicated airspeed on the co-pilot's side. Assuming this was noticed, the captain's instruments would still be functioning properly and maybe the assumption was made that a bit of water in the starboard system had frozen up.

So far so good, but then the pitot tube dropped clean away. Zero speed would definitely have been recorded before the co-pilot but as the tube fell away the stall detector would have lowered the nose flaps. The first the pilots might have known about this was when the warning lights came on, but what would they have made of them? At Mach 0.94 and 52,000ft, a Victor Mk 1 would have sailed along very comfortably like a knife through butter, but in similar circumstances a Victor 2 buffeted and shuddered around most uncomfortably because of its larger wings and bigger engine air intakes. XH668 would certainly have been bucking and rearing round the stops as its crew explored the steep turn buffet boundary, and in such circumstances the warning lights and lowering nose flaps would only have confused matters because they would have been activated for no reason as far as the crew was concerned.

Faced with stall warning indications, the natural reaction would have been to lower the aircraft's nose. Unfortunately, these distractions would have masked the most sinister 'gotcha'. At Mach 0.95 the Mach trimmer would have been fully out, but on receipt of a spurious low-speed signal it would have steadily lowered the elevators and pushed down the aircraft's nose which was very efficient aerodynamically. Moreover, the elevators would be lowering through almost 5° at a time when the pilot might have been pushing the stick forward himself. The combined nose down movement — and it might have been increased if the co-pilot had instinctively reacted to an apparent fall in his airspeed indicator before realising what was happening — would have pushed the Victor quickly beyond human recovery. It is most likely that the throttles stayed open and the airbrakes closed because both pilots were concentrating all their hands and efforts on trying to overcome the forces opposing the aileron jacks to level the wings as an essential preliminary to recovering from the dive. It would have been a futile effort and the crew was doomed as the spiral tightened. On deciding that the wings were now at a low incidence, the stall detector should have sent a signal to raise the nose flaps, but the aerodynamic forces experienced in the high-speed dive would have resisted their operating jacks as well.

Although the sequence of events in the critical period could never be established precisely, XH668's nose flaps were found to be down and its Mach trimmer actuator virtually fully retracted when it hit the sea. The jigsaw fitted together to provide all the ingredients of a major disaster. To prevent recurrence, the mountings of the pitot tubes of all Victor aircraft were modified such that the collets were locked permanently and therefore could not vibrate loose in future under any circumstances.

Right at the bottom of St George's Channel, a Victor 2 pitot head lies buried in the sand and is ever likely to remain so. And 30% of XH668 was still missing by the time Farnborough called the trawlers off on 19 November 1960, but Dr Walker's men had the answer and the winter storm season was rapidly approaching. As they sailed back into Milford Haven for the last time on Operation 'Victor Search', the 16 trawlers concluded one of the most remarkable salvage operations ever attempted and completed successfully. The

Right:
An example of what can be recovered from the deep. When Hawk T1 XX298 had to be abandoned because of a control restriction over Tremadoc Bay, Wales, in October 1984, about 70% was salvaged from the sea bed to help determine the cause.

wreckage area has been described as the bleakest and roughest patch of sea anywhere in the vicinity of the British Isles, and it was no mean feat to pick up minutiae 400ft below the surface at the end of half-a-mile or so of cable subjected to the effects of wind, tide, currents and drag. In all 1,480 men and 40 vessels, or twice the number of ships as were involved in the search for the BOAC Comet that went down off Elba in the Mediterranean six years earlier,

took part in the quest for XH668 over 15 months,. After spending £2 million at 1960 prices (the cost of a new Victor 2), Britain not only restored the bomber's credibility but also gained the record for the greatest overt salvage operation in aviation history.

Apart from being a stirring yarn of man against the elements, does the saga of the lost Victor 2 prototype hold any relevance today? Given that all accident causes must be determined satisfactorily if faith is to be retained in an aircraft or its operating procedures, it helps if the investigators can pinpoint wreckage expeditiously. All military aircraft should therefore carry underwater locator beacons. Such a beacon stays with the wreck and sends out sonar signals to guide searchers to it. But if the aircraft disintegrates in the descent, which bit should the locator beacon stay with?

The answer is the Accident Data Recorder (ADR). XH668 carried a primitive wire recorder during its test flight and some 2,500ft of wire was recovered in what first appeared to be a hopelessly tangled and broken mass. After hundreds of man hours had been spent disentangling and joining pieces together, it was established that the wire came from an unused spool. Technology has progressed apace since 1959: modern ADRs not only record voice inputs from the crew as well as other noise from an area microphone, but also take down crucial parameters such as height, speed, acceleration forces, control surface angles, pitch, roll and yaw information and engine power output. The ADR must take this down during all manoeuvres, severe turbulence, buffeting, stall-spin conditions and severe loadings. Sampling rate should be up to 256 samples/sec on a complex aircraft for at least an hour, and the total data package should be capable of being easily 'milked' by a computer. ADRs for all military aeroplanes except perhaps primary trainers should be a must, and it goes without saying that such 'black boxes' should be crash-protected.

The only record of what went on inside XH668 that fateful day was a fragment of the crew's last conversation transmitted in error and picked up purely by chance at Boscombe. It was a very weak and mangled transmission and by a freak of radio it told the inquiry that the crew was listening to 'Mrs Dale's Diary' as the aircraft went down. All very interesting but it progressed the inquiry not one jot: how much better it would have been if they had carried an ADR.

But finding out what went wrong always appears to be a negative exercise. It is much better to stop the accident happening in the first place, and the loss of such a fine aircraft as Victor XH668 showed that venturing beyond the boundaries of an aircraft's flight envelope is not to be recommended and can have unpredictable results, even for seasoned test pilots. Equally important, there are still too many flying accidents that are caused by silly little things such as a loose spanner jamming a control run, a carelessly dropped bulldog clip sucked into an engine intake or a decimal point put in the wrong place when calculating turbine blade clearances. If the loss of XH668 teaches anything, it is that everybody involved in building, operating or maintaining aircraft, no matter how remote from the flight line the task, must always give it their utmost care. Quality control and effective supervision are not just high tech preserves: they apply right across to the chap with the paint brush.

146

Loose Articles Can Kill

Although aircraft have undoubtedly become more dependable and forgiving over the years, many of the perils that have perennially challenged the pilot still remain. For example, silly hazards like loose articles have screwed up the most advanced technology for years. At the end of 1917, No 43 Squadron was deployed to Auchel when three Spads flew in, led by one of the best French 'aces', Geoffrey Comte de la Tour. The French valued their Spads highly but No 43 Squadron did not rate much to the machines' aerobatic ability and so de la Tour promised to prove his aircraft's worth. The whole squadron assembled the next day to witness de la Tour climbing into his Spad, complete with a little black leather handbag which every French pilot seemed to carry but whose contents no Briton ever managed to discover. De la Tour took his Spad up to 2,000ft and at the conclusion of a splendid exhibition of skilled piloting, he put his machine into a spin at around 2,000ft. At 1,200ft the rotation slowed up and de la Tour could be seen trying desperately to recover, but although he twice succeeded in bringing the nose up a bit from its vertical downwards position, his Spad eventually crashed through the telegraph wires at the end of the aerodrome. De la Tour died in the flames, and all because his little bag slipped and jammed the control column at its base.

To show that no matter how things change, the more they can remain the same, on 22 October 1987 the sixth production Harrier GR5, XZ325, took-off from its maker's airfield at Dunsfold. It was near the end of a series of production test flights in preparation for handover to the RAF, and the sortie was intended to clear outstanding test schedule items primarily concerning the oxygen system.

The Harrier lifted off at 16.59hrs and climbed as planned. The pilot checked in with London Military air traffic at 17.06hrs and his aircraft was seen on radar to climb to 30,000ft. It was scheduled to remain at this level for about 15min while the pilot checked the oxygen system, and the radar height readout remained constant while the Harrier tracked steadily west. At 17.33hrs, London Military tried to re-establish radio contact but there was no response to any call. A search and rescue aircraft was then scrambled and a westbound USAF C-5 Galaxy was vectored towards the Harrier, intercepting it about 140nm west of the southern tip of Eire. The Galaxy crew reported that the Harrier was pilotless and they remained in contact with it for a further 260nm during which they took still photographs and made a video recording of the ghostly sight. At 19.03hrs, when presumably out of fuel, the Harrier entered a descending spiral, disappearing into the sea 500 miles west of the southern tip of Eire.

An extensive search and rescue operation was mounted along the Harrier's track and on the evening of 23 October, the pilot's body was found by a gamekeeper about five miles west of Boscombe Down. The pilot's very damaged parachute was still attached and his oxygen mask hose was disconnected from his Personal Equipment Connector. Photos taken by the Galaxy crew showed the ejection seat still in the aircraft and the canopy frame in the closed position. Despite an intensive search, the wreck of XZ325 was never found.

Accident investigators established that the experienced Harrier test pilot was fit, alert and well-prepared for the sortie, and that he had an above average knowledge of the ejection seat and oxygen system. The fact that the unoccupied aircraft flew on for about 2hr indicated that its main systems were functioning normally other than perhaps the oxygen supply. There had been no emergency signals from the Harrier, so whatever happened must have been pretty instantaneous.

The available evidence showed that the pilot's seat harness had been released, the Parachute Deployment Rocket had fired through the aircraft canopy, and that the pilot's main parachute had then deployed pulling him from the cockpit. The opening shock plus tearing as it snagged on the aircraft structure caused serious damage to the parachute, thereby rendering it useless in retarding his fall. The pilot was killed on impact with the ground. Back-plotting showed that the pilot left the aircraft approximately 1¾min along track after making his last radio contact.

The ejection seat in use that day was a new type that, as yet, had only been fitted to the Harrier GR5. It used gas pressure, generated by cartridges, to sequence many of the required events during ejection. To meet the known facts, the 'Harness Release' gas circuit must have been pressurised. In the absence of the most crucial piece of material evidence – the ejection seat – the most probable cause of the accident was assessed as percussion firing of the manual separation cartridge.

Normally the pilot is released automatically from his seat after it leaves the aircraft, but the Manual Override capability was there just in case there was a double failure of the automatic system. Operation of the Manual Override handle, which was mounted on the righthand side of the seat-pan, fired the manual separation cartridge which pressurised the harness release circuit, freeing the pilot from his seat and deploying his main parachute. Manual separation was designed for use after the seat left the aircraft and it was safeguarded by a mechanical interlock to physically prevent operation of the Manual Override handle until the main seat firing handle had been pulled.

Given all the safety factors built into the Harrier's escape system, something weird must have happened to cause the pilot to be plucked out of his cockpit so catastrophically. All reasons for the firing of the manual separation cartridge were examined, and three possible causes emerged. Firstly, because the pilot's oxygen mask hose was found disconnected, the pilot could have become hypoxic. This lack of oxygen to the brain would have led to the pilot becoming confused and following cockpit depressurisation shortly after top of climb – it was part of the test schedule – he might have tried to eject. If the seat had failed to fire, the pilot might have pulled the Manual Override handle in a last ditch attempt to leave the aircraft. However, this sequence of events was considered very unlikely. It presupposed that the pilot did not connect the hose at all, or that it became disconnected during flight, but in either case certain warnings would have been present. All pilots are trained to check their oxygen supply

148

carefully and this test pilot, who had an above average knowledge of his oxygen system, would have had to ignore both his routine cockpit drills and the primary constituent of the sortie which was to monitor the performance of the oxygen system. There was no evidence of mask hose damage due to flailing, and therefore it was most likely that the hose became disconnected as a consequence of the accident – probably on violent impact with the ground – and not before.

The second hypothesis was that because of a main oxygen system failure, the pilot tried to select emergency oxygen manually. The Manual Oxygen lever was similar in appearance to the Manual Override handle and, although located on the left-hand side of the seat pan, the pilot could have pulled it by mistake. If the Manual Override interlock had been set incorrectly, the manual separation cartridge would have fired. However, this cause would have required three independent faults or errors and was therefore very unlikely.

The third possibility was that a loose article became lodged under the Manual Override operating rod on the right hand side of the seat: this rod connected the Manual Override handle to the cartridge firing unit sear. If the ejection seat had then been lowered, a foreign object between the side of the seat and the cockpit wall might have distorted the rod, withdrawn the sear and then fired the cartridge. On this particular sortie, the Harrier's track was close to the sun's azimuth. When the pilot levelled off at top of climb, he would have been almost facing the glare of the low sun. This glare would have hindered him from seeing items on the instrument panel, such as the warning system which he needed to observe during the test schedule. With the autopilot connected, it is wholly conceivable that he then lowered his seat to get a better view of the cockpit instruments and controls.

Subsequent tests confirmed that the Manual Override sear could be removed by a loose article and that the cockpit wander lamp – a small torch at the end of a power cable – was of a size and strength to produce the necessary effect if it had fallen from its bracket. Although no one will ever know for sure, the 'loose article' theory is the most likely cause of this tragic accident because it fitted both the known facts and was a single abnormal condition, whereas possible alternatives required a multiplicity of simultaneous errors or malfunctions. Once the Manual Override handle-to sear linkage was physically protected against inadvertent operation, the integrity of the Manual Override interlock was checked, and the offending wander lamp mounting was modified, development flying of the Harrier GR5 was safely resumed on 18 February 1988. The tragic loss of XZ325 and its pilot only went to prove that there are some accidents that no human aviator could have done anything about.

Yet that does not mean that there is nothing that a pilot can do to minimise the perennial risks. He can start stacking the odds in his favour by little things like not pushing his groundcrew to cut corners. He should rely on his engineers' advice and accept that a pilot faced with not flying is not always the best judge about what risks he should take to get his wheels off the ground. Luck as well as ability can occasionally be needed to survive, but just as the winds tend to favour the best navigator, so luck does tend to be with those who are best prepared. And if things do go wrong, aircrew must get out in time. As Samuel Butler wrote in the 17th century; 'for those that fly may fight again, which he can never do that's slain'.

The Lonely Sky

No self-respecting navigator is ever lost — at worst he becomes uncertain of his position. Yet life is always a lot easier when crossing terrain that is rich in distinctive natural and man-made features; it is over the barren areas of the earth that the aviator truly earns his money.

Every day the dry desert sun rises over eastern Libya, and in the late spring of 1943 it shone particularly on the US 9th 'Desert' Air Force's 376th Bomb Group. The 9th had recently moved up to airfields around Benghazi following the surrender of Axis forces in North Africa. Over the next few months the 9th would help pave the way for the forthcoming invasion of Sicily by attacking the enemy's air strength and his seaborne means of supply.

The 376th Group operated Consolidated B-24 Liberators and into the midst of its 514th Squadron came six new Liberators from Morrison Field, Florida, flown out via the South Atlantic ferry route. One of the six crews to arrive at Soluch-Benghazi aerodrome was led by 1-Lt William J. Hatton, a native of New Jersey and something of an 'old man' in that he was all of 25 years old on graduating from flight training less than a year earlier.

Acting as co-pilot, Hatton was assigned to fly a familiarisation trip against Palermo harbour on 2 April. In the event the mission had to turn back early because of bad weather, but the 376th was so short of crews that this abbreviated trip was considered a sufficient check of the whole Hatton crew.

Two days later they were thrown into action for the first time as part of a 25-ship high-altitude raid on Naples. The Liberators were expected to transit out at 25,000ft in broad daylight, hit the target exactly at sunset, and then break formation to come home singly under cover of darkness. As the aircraft they had ferried so carefully from Florida was otherwise engaged, the Hatton crew was given another B-24 christened *Lady Be Good*.

All nine members of the Hatton crew were at *Lady Be Good's* parking spot by 12.00hrs. A rising wind started to lift the desert sand as the ground crew took off the engine covers at 12.45hrs and by the time the 25 bombers started engines, ground visibility was down to zero. Sand whipped up by both wind and 100 whirling propellers got into every aircraft system and not a few other orifices. Nevertheless the raid had to continue: it was the last in a four-punch attack on Naples and there was no way that the 376th could be seen to leave the field wholly to Flying Fortresses.

Take-off rolls began at 13.30hrs with the first 12 aircraft forming up into Section A and heading north. The remaining 13 gulped down a double dose of sand and got airborne behind them. *Lady Be Good*, painted a dusky shade of sandblasted pink and bearing the number '64', was 21st off but as Section B approached Sardinia at approximately 200mph, six of the formation of 13 had already dropped out with sand-induced mechanical troubles. By 19.25hrs a further three more had fallen by the wayside, leaving only four Liberators to overfly Sorento, some 16 miles south of Naples. At this point the sun had already been down for 15min. It would be dark over Naples by the time they got there and as the city itself was a forbidden target, the Liberator crews decided to wheel their aircraft south and abort the mission. The small group then split up, one aircraft landing short of fuel in Malta and the other three making their way back to Libya separately, dropping their bombs en route either over the sea or on a Sicilian airfield.

At 00.12hrs the Radio Direction Finder (RDF) station at Benina, midway between Benghazi and Soluch, received a call from *Lady Be Good* asking for a bearing. The Benina operators cranked their loop aerial around until the voice count faded out; they then read '330°' off their scale and passed this bearing back to Hatton. The bearing was acknowledged and *Lady Be Good* went on her way, the crew unaware that they were being sent out into the Libyan desert.

The trouble with a single RDF station, with its solitary rotating loop antenna, was that although it could plot the aeroplane in a certain direction, it had no way of telling whether that aircraft was inbound or outbound from the station. The 'voice fade' told the operator only that the aircraft was along a certain straight line passing through the antenna. In the middle of the night on 4/5 April 1943, both the Benina operator and the Hatton crew assumed that aircraft No 64 was *inbound* to the airfield. Had there been a second RDF station sufficiently far away, it would have been possible to get a cross bearing to accurately pinpoint the Liberator's position, but no such refinement was available. All concerned had fallen into the trap of 'reading off the back of the loop'.

Unfortunately Hatton's navigator, Lt D. Hays, had plotted a strong tail wind on the outbound leg up to Naples; the crew therefore expected to meet a strong headwind on their way home. They had no way of determining the correct wind effect at night without shooting the stars as the old seafarers used to do, but astro-navigation took a lot of time and effort. Why go to all that trouble when in a few minutes the Liberator would be close enough to Benina to tune in the radio compass and follow the needle home?

The actual bearing from *Lady Be Good* to Benina at 00.12hrs was 150° — the exact opposite of what the station and Hatton thought it was. If the crew had tuned in their radio compass when they received their '330' bearing, or at any time within the previous half-hour, they would almost certainly have picked up the Soluch beacon. The Liberator's radio compass, which was working properly, would have pointed towards the airfield as the B-24 flew in, swung around as the bomber passed overhead and would have pointed directly behind as the aircraft carried on its way. But the crew were so confident of their position that they did not switch on the radio compass in time.

Hearing from Benina what they expected to hear — bearing 330° from the station, 150° to the station — Hatton or his co-pilot, Lt Robert Toner, would have waited a few minutes and then probably started a descent to be in a position to spot the coastline as the B-24 crossed it. In the event, the coast never appeared. By 01.00hrs the crew must have been getting uneasy. The radio compass would have come on but despite everyone's efforts at tuning it in, the Liberator would have been long out of range of the low-powered beacon. Because the wind had swung through 180° and therefore nothing had impeded *Lady Be Good's* return progress, the B-24 by now was at least 200 miles deep into the desert.

To the end the crew probably thought they were over the Mediterranean and that in just another few minutes they would sight the coast. They may even have tried turning 20° port just in case they had drifted off course a little to the west over the Gulf of Sirte. *Lady Be Good* had been fuelled up for a 12hr trip and once the fuel gauges indicated perilously low, everybody would have grabbed a Mae West and strapped into a parachute. When No 1 engine wound down, the seven non-pilots probably leaped out: by the time the red light came on No 3 engine, the two pilots were ready to do likewise. As they floated down, the Hatton crew would have inflated their Mae Wests ready to hit the warm sea. They could not be too far from the coast now and surely Group would launch search planes for them at daylight. It was only the instant that nine pairs of feet touched sand instead of sea that realisation must have dawned that something had gone terribly wrong. At altitude, *Lady Be Good* had had a strong tail wind instead of a head wind such that she had over-flown Benina-Soluch even before her crew started looking for it. All became clear now, but it was far too late.

The last hope for the Hatton crew, as they parachuted down into an area of desert 350 miles south of Benghazi where the daytime temperature reached 130°, was that a search party would try to find them. As dawn broke, everyone assumed that the Hatton crew had ditched and so it was towards the sea that the search effort was directed. Not surprisingly it proved fruitless, leading to the conclusion that *Lady Be Good* had most likely been shot down further out over the Mediterranean. Perhaps the crew were now prisoners. Either way, no one at Soluch ever added two and two together to realise that as the other aircraft from Section B had all landed by 23.10hrs, when Hatton broke radio silence to call for an RDF bearing 1hr 2min later, his aircraft had already passed overhead Soluch on a heading that would take it deep into the desert. It was much easier to say that the Hatton crew must have become casualties of the war while returning in the dark from their first operational mission. It would not have been the first time, and anyway there was still a war to be won. Shortly afterwards 376th moved on to a new airfield and Hatton's Liberator was forgotten.

It was a British Petroleum prospecting team who first found the wreck of *Lady Be Good* in November 1958. The pilot of the C-47 reporting the sighting did not expect it to arouse much interest: the desert was littered with decaying monuments to battles fought a decade-and-a-half earlier. However, the twisted and broken B-24 was a landmark in a bleak wasteland and its position was worth plotting.

The first men to step inside *Lady Be Good's* oven-like fuselage in 16 years arrived in February 1959. Still there was some brackish water, a flask of coffee, and the navigator's log with its last pencilled entries dated 4 April 1943. As they walked round the aircraft, the geologists saw no signs of any bullet holes or damage other than that which would have been caused by the crash. It was just as if the crew had been plucked from their aeroplane.

Extensive searches eventually revealed marker triangles left behind by the B-24 crew as they moved north, but then the trail came to an abrupt halt. It was not until February 1960 that the oil prospectors stumbled on to co-pilot Robert Toner's diary. In sparse detail it revealed that eight of the crew had joined up with each other before heading off on Monday 5 April to try and make it back to civilisation. They carried only one flask of water, some food and their parachute canopies for shelter, and on Wednesday their water was down to a good swallow each per day. Even so by the next day they had trudged 70 miles from where they had banded together. The only trouble was that their efforts had still only brought them to the Libyan Sand Sea.

By Friday 9 April, only three men had the strength left to carry on. The remainder huddled together waiting for their return or the signs and sounds of that rescue plane, neither of which ever came. Toner's diary stopped with the entry: 'Monday, April 11, 1943: No help yet, very cold night.' The little cluster of five bodies was found on an undulation in the dunes. It was months later before two more bodies came to light, the flight engineer having dragged his weary body an amazing 20 miles further on. The last resting place of the third man was never found.

Below left:
The route flown by *Lady Be Good* from Naples, passing directly over its home base at Soluch, which ended in the Libyan desert some 450 miles in the opposite direction. The dotted line shows the route the crew may have thought they were flying into a strong headwind.

Below right:
The Sand Sea of Calanshio showing how the crew of *Lady Be Good* tried to make their way out of the desert. If only they had not followed an Italian track which led nowhere, but rather had chosen to follow a British-made track leading to Zighen Oasis, they might well have survived.

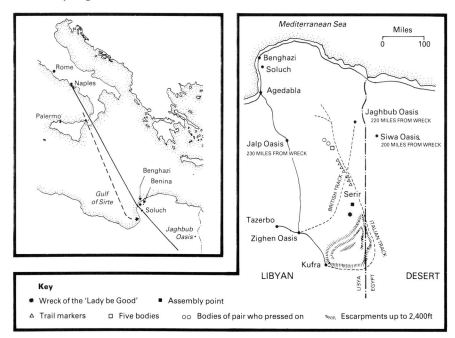

The *Lady Be Good* as discovered in the Libyan desert 16 years after crash-landing there. Although damaged on impact, the B-24 was otherwise almost perfectly preserved. *Don Livingstone*

It was a sad tale because all the dead men had made superhuman efforts to extricate themselves from a situation that they should never have got into in the first place. Everybody else made it back from Naples, why not the Hatton crew? After all, they had flown a B-24 from the States without any bother. Hatton was also the same age as the group commander, and Toner a year older.

All of which was probably their undoing. Having flown from Florida to Libya, perhaps they thought that Naples to Benghazi would be a milk run, overlooking the fact that they no longer had a protective and guiding formation to cling to or that it could be very lonely keeping radio silence at night. The Hatton crew were only 'freshmen' after all, and perhaps someone higher up in the squadron was remiss in not briefing them thoroughly enough on the pitfalls of operational flying over North Africa. A wise supervisor must ensure that his fliers learn from the mistakes of others because it is not possible for them to live long enough to make them all themselves.

But in the final analysis, the Hatton crew was never asked to do more than it was trained to do. If only they had plotted their position by the stars they would have found out that a tailwind was pushing them home much quicker than expected. Perhaps having survived their baptism of fire, too many crew members switched off mentally as they left Italian airspace. Whatever the cause, *Lady Be Good* was not the first aircraft to be run out of fuel and it would not be the last.

On the night of 11 January 1952, a C-47 took-off on a routine night instrument training mission from Eielson AFB, Alaska. On board were Capt Harvey S. Tilton (1st pilot), Capt Ernie Walker (co-pilot), Sgt Charles Medina (flight engineer), Airman 2nd Class Oscar Provencher (radio operator) and Airman 2nd Class Glen Mellon (student radio operator). Taking-off in light snow at 18.18hrs with enough fuel on board for an 8hr flight, Capt Tilton began climbing to 11,000ft while heading for Fairbanks. His radio compass was tuned to Fairbanks and when the needle swung down to show that the C-47 was passing overhead, Tilton turned towards Umiat. The deflection on the radio compass needle showed that there was a strong wind blowing, but it did not take long to set up a good wind correction angle.

Although the Gooney Bird's cabin was warm and comfortable, it was anything but quiet. Against the background din of engines and wind, Tilton and Walker would have been listening to a Morse code 'A' (dit-da) or an 'N' (da-dit). These Morse signals were the audible part of the Radio Ranging System which was the primary means of navigating long airways legs in those days. Basically, radio ranging consisted of four beacons positioned in a square. One pair of beacons in opposing sectors pushed out the Morse letter 'A' while the other pair transmitted 'N'. Where the sectors overlapped slightly, a wedge of around 3° was formed along which the signals of adjacent sectors were heard as a continuous tone. Maintaining the continuous tone kept crews on the airways centre line, and the sign of a skilful pilot in northern US and Canadian airways was one who kept a steady 'on beam' signal coming through his earphones from one Radio Range beacon to another. It was not an easy task, especially as crews had no means of accurately determining their drift or position over the ground, but Radio Ranging was the only information Tilton felt he could depend on for guidance that night.

The flight continued uneventfully until shortly after the C-47 passed over Bettles. Tilton then reported that he was experiencing a 90° wind correction because of the storm blowing outside, and that he was flying back to Bettles to reorient himself on the southeast leg of the Bettles Radio Range. Three-and-a-half hours later, with over half his fuel gone, Tilton admitted that he was lost and called for a DF steer.

Ladd Field responded. They asked the C-47 to turn twice so it could be positively identified, and then they gave Tilton a vector of 211° magnetic to bring him back to Eielson. But Capt Tilton did not acknowledge these instructions because he was confused. He was listening to the one aid on which he thought he could always rely, the dots and dashes of the Radio Ranging System. He was sure that he was on the north leg of the Northway Radio Range which told him he was 150 miles south of Eielson. Why did Ladd Field advise him to fly 211° which would surely only take him further south into the mountains?

Actually, the Tilton crew were 150 miles *north* of Eielson so the 211° steer would have brought them home. The question to be answered therefore is why did two World War 2 combat veterans like Tilton and Walker get so lost? There were many good theories, all revolving round the fact that navigation by compass in extreme northern latitudes is far from easy at the best of times. Proximity to the magnetic north pole meant that all manner of corrections had to be applied while navigating, but crews in Alaska were trained to take all these factors in their stride. It was no text book navigation error that drove the Tilton crew 300 miles off course — it was an administrative error!

On or around 11 December 1951, ground maintenance crews had swung the Bettles Radio north leg 9° counterclockwise from 156° to 147° magnetic.

Magnetic variation at Bettles was −28° which moved the 180° true bearing into a different quadrant. In practical terms, this meant that Tilton and Walker would hear da-dit when they expected to receive dit-da. The real tragedy was that the message notifying this change did not reach Eielson Operations so Tilton and Walker never stood a chance of knowing that the instrument in which they had so much faith, although working perfectly, was leading them in the opposite direction from their intended path.

The C-47's compass was only good in straight and level, unaccelerated flight, so with a 50kt west wind bouncing the aircraft around it is not surprising that the compass told the pilots one thing, the radio compass another and the radio

Below:
This map of the last flight path of Capt Harvey Tilton's C-47 on 11 January 1952 shows how confused the crew must have been. Notice the wide variation between where they were *supposed* to be, where they *thought* they were and where they *actually* ended up.

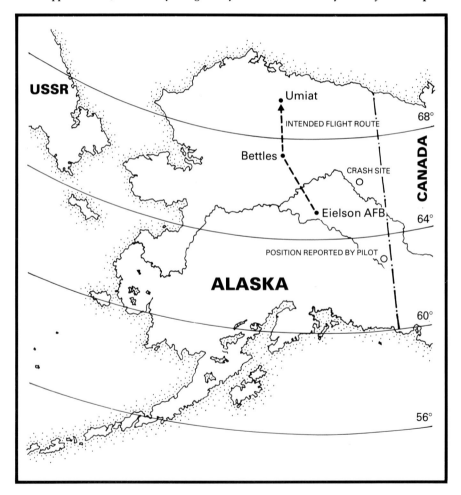

range a third. Now the ground controllers were advising Tilton to steer in a direction that he was convinced had to be wrong. We do not know what course the pilots then decided to follow, but after 6hr airborne the crew was hopelessly lost. Around midnight, Capt Tilton must have heard someone on VHF radio and put out his last transmission, 'Hearing you very weak. Do not know where I am. Receiving a strong N quad signal'.

About an hour after this transmission, Tilton must have given the order to abandon ship because the C-47 would run out of fuel within minutes. The survival aids and Gibson Girl portable radio went first, followed by the radio operators and co-pilot. As captain, Tilton was last to go and as he climbed out of the left seat to make his way aft to the exit door, what must he have been thinking? Here he was in a nice warm aircraft, which was still flying because there was nothing wrong with it: even the radios were still receiving. But there was no time for second thoughts and Tilton threw himself out to join his crew in the murderously cold Arctic.

For three more days the storm raged but nevertheless at least four members of the Tilton crew joined together in the snow. While they put their Arctic survival training to good use, the rescue crews could only wait until 14 January for the storm to break. In all, 199 sorties were then flown by dozens of aircraft from as far away as Anchorage. They divided their efforts between where Ladd Field had last fixed the C-47 north of base, and south of base where Tilton had indicated he was. Ironically, a rescue plane became lost in daylight in clear weather. The pilot reported he was getting a strong signal from Bettles Radio and only then did it dawn on all concerned that Bettles had been swung 9°.

Fifteen days after the crash, the active search for C-47, tail No 43-16249A, was terminated and the crew presumed dead. The quest was resumed when summer arrived, and after seven days the pilot of a SA-16 amphibian spotted the red-tailed C-47 some 50 miles southeast of Fort Yukon. Rescuers could only reach the area by foot, and 2½ miles behind the aircraft they found two shelters. Inside one were the bodies of Capt Tilton and the enlisted members of his crew. They were fully clothed and among the items around the bodies was a log from Capt Tilton addressed to his commander, Col Moore, setting down the conflicting evidence Tilton had faced in the air. The unfortunate aircraft captain died never understanding what had gone wrong. All he could say in conclusion was, 'Send things to my wife. I'm sorry Col. Goodbye.' Even more tragically, some 165yd south of the shelter the search team found the Gibson Girl survival radio hung high in a tree. That radio could have made contact with the searching rescuers six months earlier and it appears that the crew had found it, but they could not get it down to the ground. The remains of co-pilot Ernie Walker were never found.

There are many similarities between the crash of '249' and *Lady Be Good*, but the fact that both occurred over 30 years ago in a less technologically sophisticated age should not blind us to the chances of repetition. Modern equivalents of the C-47 — huge transports that straddle the globe — carry all the lastest navigational gadgetry. On one such flight recently, the aircraft was flying over a sparse part of Asia. 'Some time later,' reported the captain, 'the air traffic controller advised us that we were 30km to the left of track and ordered an immediate 20° alteration of course. There are few navigation aids in the region and the radio compass beacons are useless unless you are quite close to them. I took a radar fix off a well defined bay of a large lake which confirmed that we were indeed a long way off track. But why? All the intertial navigation system (INS) sets showed us to be on track. But of course they would! . . . They were in "triple mix".

'There was no time to identify and sort out the offending INS set. The first priority was to get the aeroplane back on track and keep it there. I put the aircraft back on track before the last of the lake disappeared off the radar screen and then steered the magnetic track of the airway, applying the drift shown on the INS. When I looked up from the chart some minutes later I noted that the co-pilot had re-selected INS and inserted the next way point up the track. I explained that the INS was not going to keep us on track until we had an accurate update and that the important thing for the moment was to maintain the airway by basic principles until we could discover which INS was to blame.

'The co-pilot was very unhappy with this arrangement and said so. He pointed to the INS displays which all said we were 25 miles to the right of track. I explained that the INS set relies on its present position for all its calculations. If one set is performing badly, and the sets are tripled mixed, then they will be affected. So for the moment while we were in sensitive airspace, the proper thing was to apply first principles of navigation and thus avoid sharing our immediate airspace with a foreign interceptor.

'The co-pilot remained unhappy and constantly queried my tactics, pointing to the ever increasing track error displayed. He was not finally satisfied until I requested a radar centreline check which confirmed we were on track. Later on we crossed into friendlier airspace where an accurate update was obtained. The rogue set proved to be No 3 and it was snagged at base. Had it been our sole means of navigation, I dread to think where we would have ended up.

'The whole incident made me realise that there is a generation of pilots who have not had the benefits of basic navigation and track keeping without the modern marvel of INS. The one good thing that came out of it was the co-pilot's endless questioning of my procedures when he was not satisfied with them. Perhaps the co-pilot mentioned will recognise himself in this story. If so, I hope he is still questioning decisions until he is perfectly satisfied that all is well.'

Not Positively Determined

At first glance, Georges Guynemer would have been no one's choice for France's most revered hero since Joan of Arc. Thin to the point of weediness, with more than a hint of being tubercular, he stood no chance of following his father as an honours graduate from St Cyr. But Georges' aristocratic face, and in particular his piercing eyes, well fitted the last male in a Guynemer line that stretched back at least 550 years. When World War 1 broke out Georges had to do something, and that something was to try and join the Service Aéronautique.

He was accepted, if somewhat reluctantly, as trainee mechanic because he had passed three technical courses with high marks. By dint of sheer ability and pestering, he was eventually sent off on 26 July 1915 to become a student pilot, and from then on his success in the air was immediate and phenomenal. Georges was a born aviator and even when he crashed, as he did quite often in the beginning because he pushed too far too soon, his instructor noted that Guynemer 'was not defeated and was possessed with that eternal Satanic glare of his'.

On the ground Georges was a colourless nobody, but once in the cockpit, especially of his favourite Spad *le Vieux Charles*, he became transformed into a fearless, professional killer. His attitude verged on the callous and he seemed to have a demonic compulsion to get airborne and kill Germans. Perhaps he did not expect to live long, hence the urgency of his mission. Whatever the motivation, by the end of 1916 he had become France's top ace with 25 'kills' and had received his country's highest award, the Cross of the Legion of Honour. The greater his success, the paler and more fragile-looking he became, but the combination only served to increase the adulation he received from a nation thirsting for revenge against the Hun.

There never seemed to be enough hours in the day to satisfy Guynemer's burning compulsion to fly into the battle zone. By the middle of 1917 his 'kill' tally had reached 50, but he had never looked worse in his life and there was a plan to rest Georges from combat. He would have none of it and neither would public opinion, so at the age of 23 he was promoted to command his fighter squadron in an effort to cut down his time in the air.

It was then that things started to go wrong. Faced with an increasing mountain of paperwork, and when he did find time to fly the uncertain Flanders weather made hunting next to impossible, Captain Guynemer built up a dangerous internal head of steam. Finally, on the morning of 11 September 1917, he summoned one of his friends Lt Bozon-Verdurez and told him they were going out on patrol . . . just to find trouble. The pair took-off from Villacoublay at 08.25hrs and not many minutes later they were over the shattered Belgian town of Poelcapalle. Bozon-Verdurez saw Guynemer slide down, diving *Old Charlie* after a German two-seater seen briefly through a break in the clouds. As if swallowing Guynemer up in cotton wool, the clouds closed together and Bozon-Verdurez lost contact with his leader. When Bozon-Verdurez finally broke clear, there was no sign of the Spad or the enemy two-seater. The lieutenant swept back and forth over terrain then being blasted by British artillery, but he had the sky to himself. There was nothing to do but fly home and wait.

Initially the most popular explanation was that Guynemer had stayed out on one of the solo patrols that he loved so well. After two hours had passed, and with it the limit of the Spad's ability to stay aloft, a call was made to every airfield along the line to see if Guynemer had landed elsewhere, perhaps with engine trouble. There was no trace of him on the Allied side, and if the Germans had shot down the greatest French ace, surely they would not be slow in crowing about it. But the silence remained and no sign of wrecked Spad or body was ever found.

Like Byron at Missolonghi, the reality of Guynemer's death was soon overtaken by fantasy. The Germans alleged that he died of tuberculosis and that the story of his death in action was an invention to give the romantic French a hero's death for their ace. At the other extreme, the wonderful legend was invented for French schoolchildren that Georges Guynemer 'flew so high that he could not come down again'. In so doing he became at one with the angels.

It seems almost churlish to say that hindsight now points to Guynemer bouncing a young flying officer called Kurt Wisseman. The Frenchman got within 50yd only to have his guns jam just as Wisseman thought his last moments had come. Not believing his luck, the German then turned on Guynemer and fired the fatal bullet. Elated, he wrote to his mother, 'now I need have no further fear in this war'. Eighteen days later, Wisseman lay dead in the Flanders mud.

The only other witnesses were three German infantrymen hiding from heavy artillery fire around the Poelcapalle cemetery. *Old Charlie* came down beside them and on examining the crumpled remains, they found Guynemer with a bullet through his head. The trio then turned away from the cemetery as the thunderous barrage rolled nearer, and shortly afterwards the wreck of the Spad was pulverised by shelling. No trace of aeroplane or body was ever found because, by the time the British occupied Poelcapalle, their artillery had ensured that there was none to find. The infantrymen were never traced, so they too must have perished.

Therefore there is insufficient firm evidence here to satisfy a Board of Inquiry as to the cause of Guynemer's death. We can speculate that the cause was accidental because he would not have been in any serious danger had his guns not jammed, thereby providing us pilots with one more opportunity to blame the engineers. However, as conjecture is not enough, the cause of Georges Guynemer's loss must be classed as Not Positively Determined (NPD).

Another famous NPD accident was that which proved that aircraft crashes are no respectors of rank or station. Prince George, Duke of Kent, was the fourth son of King George V and like his brothers before him, the Duke joined the Royal Navy in his teens. However, unlike the other sons of the Lord High Admiral, he found the sea disagreeable. From November 1939 he was to have been Governor-General of Australia, but then World War 2 intervened and the King agreed to the Duke joining the Welfare branch of the RAF where he was given a brief to get out and about to boost morale.

Prince George was made an Air Cdre in 1941 and in the 12 months following August 1941, he flew more than 60,000 miles on active service with the RAF. Another 1,800 miles was scheduled to be added following an inspection of RAF bases in Iceland, and on 24 August George bade farewell to his Duchess before travelling with his private secretary, acting air equerry and batman on the overnight train to Inverness. There they were met the following day by Gp Capt Francis, CO of No 4 (Coastal) OTU at Invergordon, the base on Cromarty Firth from where the Duke would fly out. On arrival at Invergordon, he was taken to meet the carefully selected crew tasked with flying him to what he termed 'the frozen north'.

The flight from Invergordon to Iceland, a distance close on 900 miles, was scheduled to take 7hr. It was to be undertaken in a standard Sunderland III, W4026, belonging to No 228 Squadron which had been flown in specially from Oban on the west coast the previous weekend. Captaining the very experienced crew was Flt Lt Frank Goyen, a 25-year-old Australian who had nearly 1,000 flying hours on ocean patrols to his credit and was sufficiently highly regarded to have flown the British ambassador to the USSR, Sir Stafford Cripps, to Moscow. Sitting alongside him was Wg Cdr T. L. Moseley, CO of No 228 Squadron and another Australian. Moseley had long service on flying boats and he was temporarily attached to the crew as a courtesy to the royal passenger. Nine further members completed the crew, comprising another pilot, a navigator, two radio operators, three gunners, a flight engineer and fitter.

On 25 August while the Duke and his party were lunching with Gp Capt Francis, the crew briefed. The weather forecast was not ideal because the whole of the British Isles was being affected by intermittent storm activity and low cloud. Nevertheless, flying was still a practical proposition because the Cromarty Firth cloud base was around 800ft and conditions were improving over the Faroes. There was no reason to delay and so the Sunderland's fuel tanks were fully filled to permit an endurance of 12hr; depth charges were also carried in case enemy submarine activity was encountered en route.

At about 12.30hrs the crew was ferried out by marine tender to W4026 where they spent the next ½hr undertaking routine pre-flight checks. The Duke was welcomed aboard at 13.00hrs and it is most likely that he and his staff went straight to the wardroom in the belly of the aircraft. Minutes later, the 1,050hp Pegasus XVIII engines were started. The Sunderland was then seen moving across the waters of the Firth, which were so calm that an unusually long take-off run was needed before the flying boat found a wave to help it lift off.

By 13.10 the aircraft was airborne and, after flying between the two precipitous Sutor rocks which marked the mouth of the Firth, it was turned northeast to follow the coastline. With a cruising speed of around 110kt, W4026 was next heard some 30min later inland from the east coast village of Berriedale. David Morrison and his son Hugh were rounding up their flock of sheep when they heard an aircraft approach from the sea, but neither father nor son saw it because of the dense mist. There followed the sound of an almighty explosion as the Sunderland crashed into a hillside and 2,500gal of aviation fuel

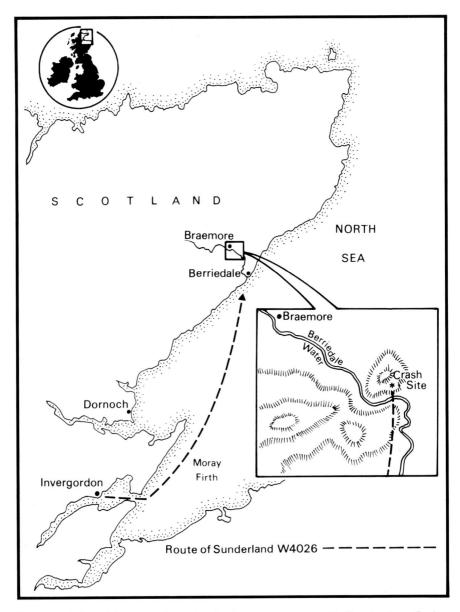

NORTH

SEA

S C O T L A N D

Braemore

Berriedale

Braemore

Berriedale Water

Crash Site

Dornoch

Moray Firth

Invergordon

Route of Sunderland W4026 — — — — — —

blew up. Of the 15 men on board only the rear gunner – Flt Sgt Andrew Jack – survived: his turret broke off on impact, hurling him burned but alive to the ground.

At the sound of the explosion, Hugh Morrison raced down the track to where he had left his motorcycle and sped into the small village of Braemore to raise the alarm. While estate workers, shepherds and crofters converged on the moors, Morrison rode on to Dunbeath to alert the local physician, 71-year-old

John Kennedy. Later that afternoon, the search parties discovered the wreckage and bodies 700ft up on a gently sloping hill which, at its western extreme, rose up to a promontory known as Eagle's Rock. Standing in the mist, Dr Kennedy and a special constable identified the Duke of Kent's body which had been flung clear of the wreckage. Despite a severe gash to the head, the Duke's features were still clearly discernible. The inscription on his identity bracelet dispelled any doubt and below his sleeve, his platinum wristwatch had stopped 32min after take-off. He was the first son of an English monarch to be killed on active service in five centuries.

The disaster made headlines around the world and on 27 August the coffin of the 39-year-old Duke, draped in a blue RAF flag, was taken from the seat of the Duke of Sutherland to the local station and thence to London. Three weeks later, King George VI drove from Balmoral to the site of his brother's death, and having seen where the flying boat came down he wrote: '. . . the ground for 200yd long and 100yd wide had been scorched by its trail and by flames. It hit one side of the slope, turned over in the air and slid down the other side on its back. The impact must have been terrific as the aircraft as an aircraft was unrecognisable when found.'

The accident occurred because the Sunderland let down as low as 700ft over high ground when it should have been much higher, or over the sea, or both. Many theories have been postulated to try and explain the discrepancy, ranging from the effects of down-draught through the influence of magnetic rocks on the compasses to plain sabotage. However, there is no evidence to support any of these suggestions.

Discounting multiple failures, the single most likely cause was that something went wrong with the aircraft's routeing. On leaving Cromarty Firth, Flt Lt Goyen and Wg Cdr Moseley had been instructed to fly northeast from Tarbat Ness across the Dornoch Firth until they met the coast again. They were then to follow the coast round, keeping over the sea all the time, until W4026 turned northwest off John O'Groats to run through the Pentland Firth towards Iceland. The main reason for not flying the quicker route directly north from Cromarty was that the heavily laden Sunderland's rate of climb was not best suited to getting above the Scottish Highlands. This profile was even less to be recommended when low cloud could just as easily surround hard as well as soft centres.

In the event, W4026 had reached 1,000ft and was still climbing as it approached Tarbat Ness. On crossing Dornoch Firth the crew would have seen the coast coming up ahead, but they must also have noted the cloud building up over it. Did Flt Lt Goyen then descend to keep in visual contact with the coastline? All that we know is that rear gunner, Flt Sgt Jack, found himself wishing that the mist enshrouding the flying boat would soon dissolve.

Evidence later came to light that it was not unusual for flying boat crews to strive for a safe altitude of 4,000ft over the Moray Firth. They then turned inland at Dunbeath to fly over the predominantly flat landscape of Caithness to the Pentland Firth, thereby saving time by cutting the corner. However, there is

163

no way that such an experienced crew, with royalty on board and squadron CO watching their every move, would have even contemplated such an unauthorised route deviation without maintaining minimum safety altitude.

Which brings us to the most salient point of all. Under the circumstances and with no pressure to cut corners, the crew would only have descended as low as they did if they felt it was safe to do so, which meant that they must have been convinced that they were over the sea. From the evidence of local witnesses, it seems likely that W4026 flew up the coast and, when around 900ft, crossed inland at a point known as The Needle just south of Berriedale Water. It then flew up the wide river valley, and perhaps the crew looking downwards through the mist saw just enough water to convince themselves that they were still over the sea. The Sunderland flew on round the 2,000ft summit of Donald's Mount, which must have been completely shrouded in mist, but then the pilot came down another 200ft, no doubt to try and keep better visual contact with what he still assumed to be the Moray Firth below. Having used up their luck missing Donald's Mount, the Sunderland crew was then struck by disaster.

On 7 October 1942, Sir Archibald Sinclair, Secretary of State for Air, reported the Court of Inquiry's findings to the House of Commons. According to Hansard,

'The Court found: first, that the accident occurred because the aircraft was flown on a track other than that indicated in the flight plan given to the pilot; secondly, that the responsibility for this serious mistake in airmanship lies with the captain of the aircraft; thirdly, that the weather encountered should have presented no difficulties to an experienced pilot; fourthly, that the examination of the propellers showed that the engines were under power when the aircraft struck the ground; and fifthly . . . That all the occupants of the aircraft were on duty at the time of the accident.'

But why did Flt Lt Goyen make such an elementary error as to keep ploughing on in such bad visibility, dropping down and down in a vain attempt to keep visual with whatever lay below? This was the action of a very green pilot, not one of his experience who knew that the only sensible thing to do was to climb on instruments above safety altitude. Wg Cdr Moseley sitting beside him would also have realised this, as would the second pilot, navigator and not a few of the seven other crew members who had a vested interest in looking after their skins. Why, less than 60 miles from base, in the middle of the day when everyone was alert and fresh, and with no pressure on anyone, did the whole flight deck crew act as accessories to the crime of flying lower than the weather conditions justified? That is what we will never know, and what makes the loss of W4026 with the Duke of Kent on board an NPD accident.

At least the wreck of W4026 was immediately available for investigation and there were some witnesses around. The really classic NPD accident usually occurs much further away from signs of life and air traffic assistance, ideally over the sea where there is little chance of any aircraft occupants surviving to tell the tale.

On 11 January 1955, five Shackleton maritime reconnaissance aircraft of No 42 Squadron RAF were detailed to carry out operational training exercises. One of the five was then to fly on to Gibraltar while the remaining four were to return to their base at St Eval in Cornwall after flights lasting between 15-18hr.

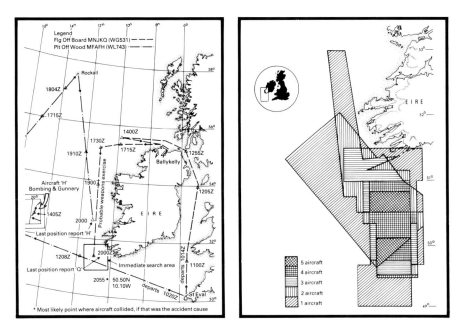

Above left:
Reconstruction of the tracks and times of the No 42 Squadron Shackletons flown by the Board and Wood crews from St Eval, Cornwall, on 11 January 1955.
Flg Off Board's crew in Shackleton 'Q' stopped off for a time at Ballykelly in Northern Ireland, but that diversion was planned and there should have been enough separation between both aircraft as they approached southern Irish waters. However, assuming that the Wood crew carried on in a straight line from their last reported position at 20.00hrs, and that the Board crew undertook a 'creeping line ahead' search position after their 20.00hrs reported position, both Shackletons could conceivably have collided 55min later around 50° 50'N/10° 10'W.

Above right:
The search pattern flown between 12-14 January 1955 to try and find the missing No 42 Squadron crews, but to no avail.

It was standard No 42 Squadron practice to leave the selection of exercise areas, routes within them, and flight plan details to the individual captains and navigators concerned. However, despite liaison between the separate navigators, three crews selected virtually the same area to the south of Eire in which to carry out tactical navigation. There was an overlapping period of 2hr when all three aircraft would be in the area at the same time.

At 10.14hrs, Flg Off George Board's crew in Shackleton Mk 2 WG531 took-off from St Eval. Six minutes later, Plt Off L. W. Wood and his crew lifted off in another Mk 2 'Shack', WL743, and both aircraft then went off to carry out a number of training exercises. Yet due to circumstances beyond their control, the Board crew had taken off 14min late. The Wood crew then left St Eval early, the combined effect being to reduce the normal half-hour separation time to 6min.

There is an old joke about a pilot who dreamed he was flying a Shackleton and then woke up to find that he was. Shackletons stayed airborne for what appeared to be an inordinately long time because, with a cruising speed well below 200kt, they could only lumber back and forth to their operating areas. Once there though, they could stay a long time on task.

During the daylight hours of 11 January, the crews of WG531 and WL743 were contacted by other No 42 Squadron aircraft on radio, and regular two-way Morse exchanges took place with St Eval. Flg Off Board's crew were due to enter the tactical navigation area off southern Ireland at 20.37hrs, with Plt Off Wood following at 21.06hrs. Their last position calls were received at 20.04 and 20.19hrs respectively, when both crews said they would call again at 21.00hrs. Plt Off Wood called some 5min early at 20.55hrs when he requested the Lundy pressure setting. These pressure settings were changed on the hour so he was told to wait for 5min, but when the ground station called him back some 3min later, there was no reply. No further contact was made with the Board or Wood crews.

Weather conditions over the whole area varied from fair with good visibility plus occasional showers in the north, to intensive rain with cloud base lowering to 300ft in the extreme south. From a dead-reckoning plot of Plt Off Wood's last position report at 20.55hrs, his aircraft was clearly in the area of good weather at the time. After every effort had been made to contact the crews or to obtain news of them, one of the other Shackletons was diverted to the north of the area. The Board and Wood aircraft carried enough fuel to stay aloft for 15hr, and so a full search and rescue alert was called at 00.45hrs. This alert got into full swing at first light, continuing until dusk on 14 January when it was finally abandoned.

With no positive evidence available, and not even an oil slick to point the way, it is impossible to come to any reliable conclusions about the cause of the loss of these two Shackletons. All we know is that both aircraft had been in regular and positive contact with base for several hours, and they both failed to pass position reports at 21.00hrs when one of them had been heard from only 5min earlier. Both aircraft were in the same area yet whatever happened to them must have been sudden and overwhelming because no distress calls were heard. Finally, despite an immediate and extensive search, no trace of any dinghy or debris was found. In fact no signs of either aircraft came to light until more than 11 years later when the starboard outer engine of Plt Off Wood's aircraft was trawled up off the southwest Irish coast, about 75 miles north of the Shackleton's last assumed position. Circumstantial evidence therefore pointed to the 'least improbable cause' of the loss of the accident being a collision between the two aircraft at approximately 20.55hrs around 50°50′N/10°10′W.

How could it happen that 18 people died in the most catastrophic of all Shackleton accidents? After all, the weather conditions were good and as there had been liaison beforehand between the crews over routeing, surely 18 pairs of eyes trained in operating and reconnoitring at low level over the sea should have been enough to keep from running into trouble? Unfortunately, that may not have been sufficient. For instance, the steady white tail light on each Shackleton was nothing like as conspicuous as modern high intensity strobe lights, which might not have been too bad had not the pilots' field of vision on the Shackleton

Above:
Shackleton Mk 2 similar to the pair from No 42 Squadron which disappeared on 11 January 1955, killing 18 with no survivors. It was the most catastrophic of all Shackleton accidents.

Mk 2 been restricted by a fairing in the forward and downward directions. Brine deposits collected on the windscreen during a long flight low over the sea would not have helped. The overall requirement to 'see and be seen' would have been affected by the fact that having already flown for 10hr when the accident probably occurred, the crews' vigilance might have been diminished.

Shackletons WG531 and WL743 were probably following each other in a practice 'creeping line ahead', a pattern flown in search and rescue missions to ensure that all parts of particular sea area were covered meticulously. This low level task was very demanding in that it imposed a special sort of strain on crew members employed as 'eyes' to look out for a capsized hull or small dinghy bobbing about amidst a mass of wave tops. The poet Swinburne described the Seamew as, 'Wide eyes that weary never, And wings that search the sea,' but humans do get weary and perhaps all those tired eyes concentrating on looking at the sea failed to see a more immediately threatening sight just up ahead until it was too late. Certainly Shackleton pilots did have a propensity to fly at exactly 1,000ft on the radar altimeter or standard pressure setting, which could have added to the chances of a collision in the midst of an occasional shower.

In the wake of this tragic accident, many improvements and changes were made in an attempt to prevent recurrence. Nevertheless, the CO of No 42 Squadron at the time, Sqn Ldr Norman Wilson, was never entirely convinced by the collision theory. From the received radio messages it appeared that his two captains had adjusted their separation to take account of the diminished time buffer, and that up to 20.00hrs that evening they were flying at the prescribed 85 miles distance from one another.

Be that as it may, midair collision must remain the 'least improbable' cause of this NPD accident. It is also worth remembering that even today, despite all the advances in technology, the majority of military midair collisions occur between aircraft in the same formation. Just because pilots and navigators talk to each other before take-off does not mean that they are inviolate thereafter. It is a sobering fact that midair collisions are not predominately random bolts from the blue — most result from poor planning, poor lookout, poor airmanship and dubious procedures or tactics.

To the trained accident investigator the NPD accident is very frustrating, but everybody else tends to relish a good *Marie Celeste*-style mystery. And once again the sea, that has shrouded strange goings-on for centuries, bears witness to the best aviation mysteries.

On 9 December 1958, Shackleton VP254 of No 205 Squadron was programmed to fly a routine anti-piracy patrol over the South China Sea. The aircraft was detached from No 205 Squadron's main base on Singapore to Labuan Island, North Borneo and on its fuselage was the code letter 'B'. Inside the Shackleton was its normal crew of 10 plus one passenger Mr A. R. Miller, a Deputy Commissioner of Police in North Borneo.

Shortly before the dawn take-off time of 05.45hrs, the crew was instructed to investigate a report from a US aircraft that a number of survivors from a shipwrecked fishing boat were stranded on an atoll 280 miles north of Labuan. At 07.10hrs the Shackleton captain, Flt Lt W. A. S. Boutell, reported that he had found 13 survivors who were flying a Chinese Nationalist flag. The aircraft then guided a fishing boat, number YF890, to the spot, and having been told at 11.57hrs to return to their anti-piracy task when the rescue was complete, the Boutell crew reported that a USAF B-50 had arrived two minutes earlier. By this time, VP254 had been airborne for just over 6hr, but then it just disappeared off the face of the earth.

At first light the following day, an extensive air and sea search was mounted over six days. Then on 15 December, as Flt Lt John Elias and his No 205 Squadron crew flew over Sin Cowe island about 12 miles northwest of VP254's

The last flight of Shackleton VP254 from Labuan to who-knows-where near Sin Cowe island.

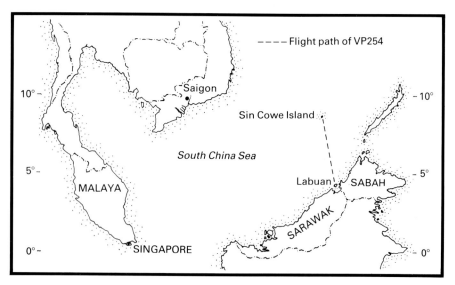

last known position, they spotted the marking 'B 205' laid out in white coral rock on a sandy beach at 09°53'N/114°20'E. A party from the Royal New Zealand Navy frigate *Rotoiti* was landed on Sin Cowe where they found an RAF officer's cap, an aircrew watch and a grave marked with a wooden cross on which 'B 205' had been carved. The body in the grave was transferred to the aircraft carrier HMS *Albion* whose helicopters later removed the simple cross. At Labuan the body was identified as that of Flt Sgt D. N. G. Dancy, the flight engineer of VP254.

No further trace of either Shackleton or crew was ever discovered. Eventually officers from HMS *Albion* traced a Taiwanese fisherman, Gan Chung-Huang, who captained a vessel named *Ray Fu Chen*. Captain Gan's crew claimed to have seen VP254 crash into the sea after passing low over them about 7 miles south of Sin Cowe island. The fishermen found only Flt Sgt Dancy's body, plus the cap and watch, and remembering the markings they had seen on the Shackleton's side, they took the body to the island, buried it in a reverent and Christian manner, and carved out the wooden cross. Guessing that a search would be made, they also laid out the markings on the beach.

Flt Sgt Dancy's body was re-interred with full military honours in the Military Cemetery at Singapore. When the British withdrew from the Far East in 1971, the last No 205 Squadron Shackleton to leave carried the cross to transfer it to St Eval parish church. Capt Gan Chung-Huang later received a letter of thanks and a £70 cash grant for himself and his crew in recognition of their efforts. No definite cause of the crash was ever recorded, and it could only be surmised that disorientation or misjudgement of height at low level over a glassy-calm sea were the most likely probabilities.

However, in March 1959 it was rumoured in Hong Kong that the Shackleton crew had not all been killed and that in fact they had buried Flt Sgt Dancy, their only casualty. A look at the chronology of events certainly implies that there was more to this story than met the eye.

In their last message at 11.57hrs, the Shackleton crew referred to a USAF B-50 coming on station concurrent with fishing vessel YF890 being guided to the Chinese Nationalist survivors. Capt Gan of the *Ray Fu Chen* estimated that the crash occurred 'shortly after noon local time' and Flt Sgt Dancy's watch had stopped at 11.55 and 16sec following a severe impact. Assuming that Dancy's watch stopped as a result of the accident, and allowing that it may have been some minutes slow, VP254 cannot have crashed far from YF890 or the B-50, yet neither reported having seen the accident.

Perhaps that was not too unlikely but what is more surprising is that the site of Flt Sgt Dancy's grave — so well marked and clearly visible from the air — was not seen before the sixth day of the search, even though it was so close to the last known position of the aircraft and therefore the focal point of the operation. *Ray Fu Chen's* captain said that his crew buried the body on the same day as the crash, so presumably the cross and other markings to draw attention to the spot were also in place by the first day of the search. Although the cross erected on Sin Cowe island bore the date '9 Dec 1958' and the inscription '13h 55m TIME' as well as the letters 'B 205', the burial could have taken place much later.

If so, what happened in the meantime? The only chance of finding out lay with the crews of the fishing boats involved. After a process of elimination, it was concluded that fishing vessel YF890 was probably Filipino, but neither craft nor crew were ever traced. On the other hand, the captain of the *Ray Fu Chen* was more than willing to talk to the British Consul in Tamsui. According to Capt Gan, he left Taiwan on 3 November and a week later arrived to fish near Ching Hsaio island (the Chinese name for Sin Cowe). On 19 November the *Ray Fu Chen* was water-logged in a storm, remaining in that condition with engines stopped and on the look-out for assistance for three weeks. On 9 December the crew witnessed the crash of a 'foreign aeroplane', so the captain launched his ship's sampan to go to its assistance. His crew recovered a body, buried it and then returned to the *Ray Fu Chen*. They drifted on for another three days until, on 12 December, they struck a submerged rock and their ship sank. The crew escaped in the sampan to an occupied island from whence they returned to Taiwan at the end of January 1959.

Yet if this story is weighed against the factual evidence, the results are surprising. On 9 December, from a trawler which had been adrift in a water-logged condition for nearly three weeks, the crew recover a body in their sampan and take it to Sin Cowe island. There, despite the presence of timber, they construct a coffin from planks 'intended to keep crew members afloat if the trawler were to sink'. They then bury the body, lay out markers to attract attention and erect an 8ft-high cross which, in the view of one expert, was sufficiently well made to have required a modicum of tools. Incised on the cross were a number of details in English, together with the RAF roundel painted in three colours, all of which were drawn by non-English speakers from the fleeting memory of VP254 passing overhead.

Perhaps even more surprising, Capt Gan and his men then re-embarked on the *Ray Fu Chen*, drifting on for another three days before coming to grief. Why did they not remain on Sin Cowe where there was fresh water in drums, where they had already made strenuous efforts to attract the attention of passing aircraft and where there was already signs of habitation? Having demonstrated that they possessed wood, tools and carpentry skills, why did they make no attempt to repair their own craft? And how is it, after apparently remaining in the same area for 20 days before the Shackleton crash without drifting more than a few miles, they immediately managed to sail their sampan northeast into the teeth of a 20kt wind as soon as VP254 went down?

It is not impossible that the Taiwanese captain embellished some of his tale. Nevertheless, we have nothing more to go on than the highly suspect nature of the evidence and some Hong Kong club gossip, a commodity not known for underplaying anything resembling a nine-days' wonder. Certainly the discrepancies do not throw any light on the mystery of why a four-engined aircraft crashed so swiftly that it did not have time to send a message. Moreover, if only one of the Shackleton crew did perish, all nine of the remainder must have been plucked away from searching eyes pretty quickly and if so, by whom — local pirates, Communist Chinese, men from Mars? Yet try as you will, the pieces of evidence do not fit together because they give the impression of belonging to more than one jigsaw. The loss of VP254 and its crew is, and is ever likely to remain, a classic NPD accident.

The Last Word

'Out of this nettle, danger, we pluck this flower, safety.' Thus wrote that clever fellow Shakespeare, who, despite living four centuries ago, caught the essence of the apparent modern contradiction between safety and military flying. Nevertheless, there may still be those who see little linkage between the accidents of yesterday and aviation as its stands today or will be tomorrow. Yet if we discount the superficial differences, there is usually an unbroken thread which runs through military aircraft accidents across the ages.

Take the airship R 38 for example, which crashed on 24 August 1921 after being subjected to steeper and steeper turns until the strain on the hull built up to a point where its structure broke up. We will never know why the captain allowed his craft to get into such a position, but happen it did.

Thirty-one years later, aeronautics had progressed to such an extent that the US National Advisory Committee for Aeronautics — NASA's predecessor — decided to expand its research aircraft programme to explore flight characteristics of atmospheric and exo-atmospheric designs capable of Mach 4 — Mach 10 at altitudes between 12 and 50 miles. To meet this specification, North American was awarded a contract to build three versions of what the USAF designated the X-15. Metal began to be cut in September 1956 and just over two years later the first X-15 was officially rolled out of North American's Los Angeles plant. The rocket-propelled high-performance research aircraft was to be air-launched from a modified Boeing B-52 carrier, and by 7 February 1961 the first X-15 had reached both 136,500ft and Mach 3.5/2,275mph.

Below:
The first X-15, at the hands of NASA test pilot John McKay, comes into land with a F-104C 'riding herd' on 28 October 1960. The X-15 was returning having clocked up Mach 2.02/1,333mph and reached 50,700ft.

X-15 flights soon provided valuable stability and control information, but much more was promised once a new liquid-fuelled rocket motor with a 57,000lb sea-level thrust rating became available. On completion of North American's X-15 flight test programme obligations, the USAF, USN and NASA embarked on full exploration of the aircraft's performance envelope. To meet the demands of the Space Age, the USAF encouraged the cream of its pilots to join its test pilots' school. There they undertook six months of test pilot training and one pilot — Maj Mike Adams — was so good that he could choose between becoming a NASA astronaut and flying the X-15. He chose the latter.

By the autumn of 1961 the X-15s had already been above 215,000ft and at the end of the year the third X-15 (serial number 56-6672) arrived back on the line after being rebuilt following an engine explosion caused by a faulty valve. During this rebuild the aircraft was fitted with a Minneapolis-Honeywell MH-96 self-adaptive flight control system. This unit made the X-15 easier to fly because an electronic device automatically fed information into the system to control the aircraft without assistance from the pilot.

Despite pushing back many frontiers during their flight test history, the safety record of the X-15s was marred by few accidents and even fewer injuries. However, on 15 November 1967, seven months after an X-15 set an unofficial world absolute speed record of Mach 6.7/4,520mph, Mike Adams set off in 56-6672 on his seventh X-15 mission and his third at the controls of the MH-96-equipped aircraft. Following an apparently normal launch, he climbed skyward and soon achieved the mission's speed and altitude objectives of Mach 5.2 and 266,000ft respectively. Part of the test schedule was wing rocking to assist optical tracking of the aircraft's exhaust plume. It was noted by ground personnel monitoring the recording instrumentation that when Adams initiated these tests, the X-15 by far exceeded the authorised bank angles.

Less than a minute later, and while at a speed in excess of Mach 5, the X-15 skewed some 90° to its ballistic flight path and entered a spin. The spin continued from 230,000ft down to 125,000ft whereupon the aircraft went into a conventional dive. Pitch oscillation then started, saturating the automatic flight control system computer and increasing the pitch control problem. Structural loads now increased to +15g, which was far beyond the X-15's design limits of +7.33 to −3g. The aircraft disintegrated and Adams, in a conventional ejection seat, did not survive.

Many months later the official accident report was completed. The cause was traced to a combination of errors and mechanical failures, not the least of which was the pilot permitting the X-15 to deviate from its flightpath. As with the experienced R 38 airship captain, we will never know why such a superb aviator as Michael Adams allowed this to happen. It could have resulted from instrument misinterpretation or cockpit distractions or vertigo, but for whatever reason the MH-96 system sustained a longitudinal pitching oscillation that allowed the control system to impose excessive flight loads on the airframe. The X-15 was a stepping stone to the stars but the loss of 56-6672 also forged a link back to the R 38. All of which goes to show that there are no new accidents, only new pilots.

Select Bibliography

H. H. Balfour, *An Airman Marches*; Hutchinson, 1933
Ralph Barker, *Survival in the Sky*; William Kimber, 1976
R. Beamont and Arthur Reed, *English Electric Canberra*; Ian Allan Ltd, 1984
R. Beamont, *English Electric P1 Lightning*; Ian Allan Ltd, 1985
Ray Braybrook, *Hunter*; Osprey, 1987
Annie Bullen and Brian Rivas, *John Derry*; William Kimber, 1982
John Chartres, *Avro Shackleton*; Ian Allan Ltd, 1985
Harry B. Combs with Martin Caidin, *Kill Devil Hills*; Secker & Warburg, 1979
Len Deighton and Arnold Schwartzman, *Airshipwreck*; Johnathan Cape, 1978
Bill Gunston, *Early Supersonic Fighters of the West*; Ian Allan Ltd, 1976
Michael Hardwick, *The World's Greatest Air Mysteries*; Odhams, 1970
Robin Higham, *The British Rigid Airship, 1908-1931*; Foulis, 1961
Bill Hooper, *The Passing of Pilot Officer Prune*; Midas Books, 1975
Paul Jackson, *The Last Flight of R-38*; Air Britain, March 1970
Douglas A. Koster, *Operation Victor Search*; Terence Dalton, 1979
Ian Mackersey, *Into the Silk*; Robert Hale, 1956
Francis K. Mason, *Hawker Hunter*; Patrick Stephens, 1981
Herbert M. Mason, *High Flew the Falcons*; J. B. Lippincott, 1965
Dennis E. McClendon, *The Lady Be Good*; Aero Publishers, 1962
Jay Miller, *The X-Planes*; Midlands Counties, 1983
Stephen M. Morrisette, *'Whatever Happened to Ernie Walker?'*; Flying Safety,
    March 1988
Heinz Nowarra, *Heinkel He 111*; Jane's, 1980
Bryan Philpott, *Meteor*; Patrick Stephens, 1986
Bryan Philpott, *English Electric/BAC Lightning*; Patrick Stephens, 1984
Arthur Reed, *BAC Lightning*; Ian Allan Ltd
Raymond L. Rimmell, *Zeppelin*; Conway Maritime Press, 1984
Douglas H. Robinson and Charles L. Keller, *'Up Ship' — US Navy Rigid Airships,
    1919-1935*; Naval Institute Press, 1982
John W. R. Taylor, *CFS: The Birthplace of Air Power*; Putnam, 1958
Christopher Warwick, *George and Marina*; Weidenfeld & Nicholson, 1988
Chuck Yeager and Leo Janis, *Yeager*; Century, 1985

Index

(Dates refer to specific accidents)